Health, Safety and Environment Legislation

A Pocket Guide

Health, Safety and Environment Legislation
A Pocket Guide

R. Day, BPharm, MPhil, LLB, MRPharmS, Barrister
AEGIS Environmental and Safety Services Ltd, Dartford

E. Rowland, MA, LLM, Barrister
Edward Rowland Consulting

Coordinating Editor

J.A. Reader, BSc, PhD, CChem, FRSC, FRSA
AEGIS Environmental and Safety Services Ltd, Dartford

RS•C
advancing the chemical sciences

ISBN 0-85404-497-3

A catalogue record for this book is available from the British Library

Published by The Royal Society of Chemistry,
Thomas Graham House, Science Park, Milton Road,
Cambridge CB4 0WF, UK
Registered Charity Number 207890

For further information see our web site at www.rsc.org

Typeset by Land & Unwin, Bugbrooke, Northamptonshire, UK
Printed by TJ International, Padstow, Cornwall

Preface

'Ignorance of the law excuses no man'

John Selden, legal antiquarian and politician, (1584–1654)

During the last decade, following the publication of the Government's White Paper, 'This Common Inheritance' in 1990, concern for the environment has continued to grow. Moreover, re-organisation of the Government's management of the environment with the establishment of the Environment Agency has been associated with a proliferation of environmental legislation. For those in business, employers and employees, it is always difficult to decide on the best ways to remain well informed on environmental legal requirements and individual responsibilities. Since ignorance of the law is no defence, familiarity with the law throughout the organisation must be encouraged. *A Pocket Guide to Environmental Law*[1] was first published in 1993 to provide compact, up to date, information in a readily accessible and easily manageable form – a 'first port of call' for environmental law. It was easily carried to provide immediate information out on site and could be used as a primary reference document in the office.

Long before care for the environment was a popular topic, concern for the health and safety of industrial workers was well established. *Redgrave's Factories Acts*[2] was first published more than 120 years ago. The legal framework for health and safety legislation is more established than that for environmental legislation; the major piece of legislation, for the former, being the Health and Safety at Work *etc*. Act 1974, and from 1974 numerous regulations have been made under the Act to keep the law up to date.

In the last few years, health and safety management and environmental management have shared more and more common ground; accordingly, following the success of the *Pocket Guide to Environmental Law*, last published in October 1997, the environmental law text provided by Edward Rowland has now been rewritten and updated by Bob Day who has also written a new section on Health & Safety Law, the vast majority of which concerns the Regulations made under the Health and Safety at

Work *etc*. Act 1974. The two closely related sections on Health & Safety and Environment are presented together in this single volume.

The enlarged guide is still intended to be an easily carried preliminary reference document which is supplementary to larger and more detailed volumes and CDs on Health & Safety and Environmental Law. The guide is not intended to give legal advice and readers should seek professional aid when dealing with specific situations.

For the Guide to retain a conveniently small size, it contains the most relevant Acts and Regulations omitting certain specific areas of limited application, for example, Health and Safety of Offshore Installations.

Each of the 98 subsections is divided into 12 parts for easy access to information:

1. The relevant Acts and/or Regulations.
2. Key dates related to that legislation.
3. The purpose of the legislation.
4. The methods by which that purpose is to be achieved.
5. The relevant regulatory authorities.
6. What and whom the legislation covers.
7. The requirements of employers/employees to meet the legislation.
8. What needs to be disclosed.
9. Any complementary legislation.
10. Related EC legislation.
11. Comments.
12. Related documentation.

The readers attention is drawn to the most recent additions; namely, information on CHIP 3, July 2002 (Section 3.4), Genetically Modified Organisms, October 2002 (Section 9.1), Control of Asbestos at Work, November 2002 (Section 3.2), COSHH, November 2002 (Section 3.6) and Control of Lead at Work, November 2002 (Section 3.8). All the material is current as of 21 November, 2002.

Finally, it pays to be proactive and not to be forever fighting a rearguard action. The relevant regulatory authorities, in my experience, have been understanding and helpful in discussing difficulties that may arise in the management of Health & Safety and Environmental topics. They have knowledge, from a wide assortment of operations, which should never be ignored.

We would like to thank Mrs Janet Freshwater CChem MRSC, Commissioning Editor, and her colleagues at The Royal Society of Chemistry, for their help and support in the publication of this guide.

J.A. Reader

1. E. Rowland, *Pocket Guide to Environmental Law*, 1st edition, 1993, ILX Publishing.
2. Redgrave's Factories Acts has undergone several changes over the years and is now available as *Redgrave's Health and Safety*, 4th edition, 2002, Butterworth & Co.

The editor would appreciate any suggestions for additions or deletions and corrigenda for a further edition.

Contents

A HEALTH AND SAFETY

1 GENERAL

1.1 General Framework

LEGISLATION

- Health and Safety at Work *etc.* Act 1974.

KEY DATES

- Duties on employers and others in force from 1/4/75. Other sections in force from 1/9/74 or 1/1/75.

PURPOSE

- To secure the health, safety and welfare of persons at work and to protect persons other than persons at work against the risk to health or safety arising from work activities.

APPROACH

- Establish general legal duty to ensure health, safety and welfare of anyone affected by work activities.
- Establish the Health and Safety Commission (HSC) and the Health and Safety Executive (HSE) to take appropriate action to further objectives of Act.
- Enables detailed Regulations and Codes of Practice to be passed under the Act.
- Provides enforcement regime – making breach a criminal offence.
- Establishes Employment Medical Advisory Service.

REGULATORY AUTHORITY

- HSE.
- Environmental Health Departments of local authorities.

COVERAGE

- Legislation applies to all work places and all persons at work, except for servants in domestic premises. Crown bodies, *e.g.* HM Forces are covered but they cannot be prosecuted under the Act.

COMPLIANCE REQUIREMENTS

- Employers must comply with a duty to ensure the health, safety and welfare of all their employees so far as is reasonably practicable. This is one of the most important sections of the Act and in essence the employer must provide a safe system of work. A non-exhaustive list of examples includes: making adequate provisions for safety with respect to plant and systems of work, articles and substances, information, instruction and training, supervision, place of work and working environment.
- A permit to work system may be an appropriate system to demonstrate compliance with the above duty in certain circumstances, *e.g.* hot work.
- Employer of five or more employees must prepare and revise a written health and safety policy including organisation and arrangements to carry out the policy. This policy and any revisions must be brought to the notice of all employees.
- Employers and the self-employed are obliged to ensure, so far as is reasonably practicable, that persons other than their employees are not exposed to risks to their health and safety.
- Persons in control of premises to ensure, so far as is reasonably practicable, that the premises are safe and without risks to health.
- Designers, manufacturers, importers and suppliers of articles and substances for use at work to ensure adequate health and safety information is available for foreseeable uses.
- Employees are required to take care of their own safety and that of others and cooperate with their employer.
- No-one must intentionally, or recklessly, interfere with safety measures and equipment.
- Employers may not charge for provision of safety measures specifically required by law.

DISCLOSURE

- Employees are required to be informed of the safety policy
- Articles and substance safety information.
- Any improvement or prohibition notices will be on a public register.

COMPLEMENTARY LEGISLATION

- Management of Health and Safety at Work Regulations 1999.
- All health and safety legislation made under the Act which is the vast majority of legislation in the Health and Safety section.

RELATED EC LEGISLATION

- Relevant EC Directives are listed under the summary of Regulations most of which are enacted by Regulations made under this Act for the Health and Safety section.

COMMENTS

- The Act establishes the HSC and HSE and specifies their composition and functions.
- The Act is enabling legislation which means provision is made for Regulations and Approved Codes of Practice to be made under the Act. The intention is to replace all health and safety piecemeal legislation in force prior to the Act with new up to date legislation.
- The Act's intention is to make health and safety as important as any other company function, *e.g.* finance. Therefore breach of the Act and associated Regulations is a criminal offence punishable by fines and/or imprisonment. Criminal liability on individual persons (within a company) can occur if breach of the legislation is committed with the consent, connivance or neglect of that person. This liability is restricted to directors, managers, company secretary or anyone purporting to act in such capacity within the company.
- The HSE and local authorities have the power to appoint inspectors under the Act. The inspectors have very wide powers

including entering premises for specified purposes, issuing improvement notices (notice specifying an act must be done in a specified time) and prohibition notices (notice prohibiting certain specified acts either indefinitely or until specified remedial measures are undertaken) and taking away certain articles.

- The Act contains powers conveyed on the HSC to require information to be disclosed to them in specified circumstances.
- The Act maintains the Employment Medical Advisory Service and specifies its functions and powers.
- The main duty under the Act is to maintain a safe system of work and this will involve looking at and checking procedures, practices *etc*. However, there was no strict legal duty to conduct formal written risk assessments until the Management of Health and Safety at Work Regulations 1992 (now re-enacted as 1999 Regulations) introduced this requirement.

RELATED DOCUMENTATION

- Safety policy.
- All documentation relating to health and safety.

1.2 General Management Framework

LEGISLATION

- Management of Health and Safety at Work Regulations 1999 (SI 1999 No. 3242).

KEY DATES

- In force from 29/12/99.

PURPOSE

- To establish more stringent requirements regarding health and safety organisation, training, health surveillance and information and to impose obligations to temporary workers, pregnant women and young persons in line with the UK's obligation to enact EC Directives.

APPROACH

- Requires employers to undertake self assessment of the risks associated with their work, to identify and implement the necessary controls, and to put in place the management structure, systems and procedures, to maintain adequate control of safety risks.

REGULATORY AUTHORITY

- HSE.
- Environmental Health Departments of local authorities.

COVERAGE

- All persons at work, except for servants in domestic premises. Crown bodies, *e.g.* HM Forces are covered but they cannot be prosecuted under the Regulations.

COMPLIANCE REQUIREMENTS

- Risk assessments to be carried out by employers and the self-employed to ascertain what they have to do to comply with obligations under health and safety legislation. Such assessments are to be reviewed if there is a significant change or there is reason to believe an assessment is no longer valid. Written risk assessments are required if the employer employs five or more employees.
- Any preventative and protective measures identified should conform to the following order of preference:
  - Avoid risk.
  - Evaluate unavoidable risks.
  - Combat risks at source.
  - Adapting work especially design equipment to alleviate certain activities, *e.g.* monotonous work.
  - Adapting to technical progress.
  - Replacing dangerous by non-dangerous or less dangerous.
  - Develop an overall prevention policy.
  - Giving collective protective measures priority over individual measures.
  - Giving appropriate instruction to employees.
- Employers must make, give effect to and in certain cases record appropriate health and safety arrangements.
- Employers must appoint an adequate number of competent persons (as defined) to enable them to comply with health and safety legislation. If there is a competent person employed by the employer that person shall be the competent person in preference to a competent person not in his employment.
- Employers shall establish, and give effect to, procedures for serious and imminent danger areas.
- Employers shall ensure that any necessary contacts with external services, *e.g.* first-aid are arranged.
- Employers are obliged to give certain specified information on health and safety to their employees including the risk assessments required under these Regulations.
- Certain specific obligations are detailed concerning employers duties towards other employees from different companies who are working on their premises.

- Employers must ensure that employees are provided with adequate health and safety training and consider individually their capabilities when entrusting tasks to them.
- Employees must use machinery, equipment, dangerous substances, means of production and any safety equipment in accordance with any relevant training and instructions.
- Employers must provide temporary workers with health and safety information prior to commencement of work.
- Employers shall ensure that employees are provided with adequate health surveillance considering the risks to their health and safety.

DISCLOSURE

- Employees have the right to know the outcome of the assessments and details of emergency procedures.
- Employees have the right to know the risks arising from contractors or other employers on the site.
- Employers must disclose risk information to other employers on site.
- Employers must provide employees with such information as is necessary to do their particular job safely.
- Health surveillance can raise clinical confidentiality issues. Employees have the right to know the results.

COMPLEMENTARY LEGISLATION

- Health and Safety at Work *etc.* Act 1974.

RELATED EC LEGISLATION

- EC Directive 89/391 on the introduction of measures to encourage improvements in the safety and health of workers at work (*the Framework Directive*).
- EC Directive 91/383 supplementing the measures to encourage improvements in the safety and health at work of workers with a fixed-duration employment relationship or a temporary employment relationship (*the Temporary Workers Directive*).

COMMENTS

- The Health and Safety at Work *etc.* Act 1974 together with these Regulations establish the framework for the management of health and safety at work.
- Section 2 Health and Safety at Work *etc.* Act 1974 provides that employers must ensure a safe system of work but these Regulations, whilst still being general in terms, expand on how to provide such a safe system *e.g.* by specifying risk assessments must be undertaken.
- The Regulations are general and more specific guidance is available in *L21 Management of Health and Safety at Work – ACOP.*
- Note: requirements concerning pregnant workers and young persons are covered separately in the Employees section.

RELATED DOCUMENTATION

- Written risk assessments.
- Employees training records.
- Health surveillance records.
- Health and Safety structure and responsibilities.
- Serious and imminent danger procedures and employee records of training in such procedures.

1.3 Enforcement

LEGISLATION

- Health and Safety (Enforcing Authority) Regulations 1998 (SI 1998 No. 494).

KEY DATES

- In force from 1/4/98.

PURPOSE

- To clarify whether HSE or local authority are responsible for enforcing health and safety laws (Health and Safety at Work *etc.* Act 1974 and other relevant statutory provisions within the meaning of this Act) in a particular premises.

APPROACH

- Allocates enforcement on the basis of the main activity on the site.
- The *local authority* will be the enforcing authority for premises where the main activity is: retail, exhibition, office activities, catering, caravan sites, consumer services in a shop, launderettes, cosmetic services, recreation, hiring of pleasure craft, animal accommodation, undertaking places of worship, airport car parking and child care facilities.
- The *HSE* is the enforcing authority in any other case and will always be the enforcing authority in specified cases of: mines and quarries, fairgrounds, broadcast, construction, work on gas or electrical systems, ionising radiation, agriculture, work on board a ship, skiing facilities, fish maggots, game breeding except in a zoo, pipelines and railway operations.
- Transfer of enforcement responsibility from HSE to local authority and *vice versa* can occur by agreement of both bodies and is subject to affected parties being informed.
- HSE, or the local authority, can apply to the HSC to determine who is responsible for enforcement at particular premises if there is uncertainty.

REGULATORY AUTHORITY

- HSC.

COVERAGE

- All work activities.

COMPLIANCE REQUIREMENTS

DISCLOSURE

COMPLEMENTARY LEGISLATION

- Health and Safety at Work *etc.* Act 1974.

RELATED EC LEGISLATION

COMMENTS

RELATED DOCUMENTATION

1.4 Insurance

LEGISLATION

- Employers' Liability (Compulsory Insurance) Act 1969.
- Employers' Liability (Compulsory Insurance) Regulations 1998 (SI 1998 No. 2573).

KEY DATES

- In force from 1/1/72.
- 1998 Regulations in force from 1/1/99 except provisions relating to certificates in force from 1/4/99.

PURPOSE

- To ensure that adequate financial resources are available to compensate employees whose employers are liable for causing bodily injury or disease to them.

APPROACH

- Imposes legal obligation on employers to take out insurance against their liability to employees for bodily injury or disease sustained during their employment.

REGULATORY AUTHORITY

- HSE.

COVERAGE

- Employed persons at work.
- Exemptions include – undertakings under national ownership, police, local authorities, health authorities and government departments.

COMPLIANCE REQUIREMENTS

- Employers must insure against their liability for bodily injury or disease sustained by their employees during their employment.
- Insurer must be authorised. Cover to be at least £5 million.
- Insurance certificate must be displayed at each work place and kept for at least 40 years.

DISCLOSURE

- Full and frank disclosure of all relevant facts to insurers.

COMPLEMENTARY LEGISLATION

- Offshore Installations and Pipeline Works (Management and Administration) Regulations 1995.

RELATED EC LEGISLATION

COMMENTS

- Normal limit of cover offered is £10 million to include legal costs.
- 1998 Regulations increased minimum level of cover from £2 million to £5 million, changed the certificate to be displayed so that it is easier for employees to understand, introduced 40-year record keeping of certificates and gave HSE inspectors the right to see past as well as present certificates.

RELATED DOCUMENTATION

- Insurance certificate, application form, policy, terms and conditions.
- Any material disclosures to insurers including insurance survey, audit *etc.* of premises covered.

2 EXTERNAL NOTIFICATIONS

2.1 Accident Reporting

LEGISLATION

- Reporting of Injuries, Diseases and Dangerous Occurrences Regulations 1995 (SI 1995 No. 3163) (RIDDOR)

KEY DATES

- In force from 1/4/96.

PURPOSE

- To establish the framework within which injuries, diseases and dangerous occurrences must be reported to the enforcing authority.

APPROACH

- Specify lists of the types of injuries, diseases and outcomes of incidents that must be reported and the method and timescale of making the report.
- Define the responsible person who should make the report and to whom it should be sent.

REGULATORY AUTHORITY

- HSE.
- Environmental Health Departments of local authorities.

COVERAGE

- Fatal and certain major non-fatal injuries arising out of, or in connection with, work.

- Specified dangerous occurrences.
- Specified reportable diseases.
- Additional provisions relate to mines, quarries and offshore workplaces.
- Responsible person, including employers and persons having control of premises in which the reportable event occurred and which are used to carry on any trade, business or other undertaking.

COMPLIANCE REQUIREMENTS

- The responsible person shall forthwith notify and within 10 days send a report to the enforcing authority if any of the following occur in connection with work:
  - Any person dies.
  - Any person suffers a specified major injury *e.g.* any amputation or loss of sight.
  - There is a specified dangerous occurrence *e.g.* the collapse of lifting machinery.
  - Where a person is incapacitated from doing his normal work for three consecutive days (not including day of incident) as a result of injury arising out of an accident.
- The responsible person shall inform the enforcing authority in writing of any death arising within one year of the accident causing the injury.
- Where a person at work suffers from one of the specified diseases and their work falls into specified categories then the responsible person must immediately send a report on form *F2508A* to the enforcing authority.
- Certain requirements for the reporting of gas incidents are detailed.
- Records must be kept of any reported deaths, dangerous occurrences, injuries or diseases for at least 3 years.

DISCLOSURE

- The enforcing authority may request further records/ information.

COMPLEMENTARY LEGISLATION

- Social Security (Claims and Payments) Regulations 1979.

RELATED EC LEGISLATION

COMMENTS

- Person required to report varies according to circumstances, type of incident, and victim.
- The Social Security (Claims and Payments) Regulations 1979 provide that certain employers must keep an accident book and record all accidents no matter how minor.
- The Regulations provide for a defence if the employer did not know of the reportable event and can demonstrate he had taken all reasonable steps to have such events brought to his notice.
- Reference should be made to L73 *A Guide to the Reporting of Injuries, Diseases and Dangerous Occurrences Regulations 1995 (RIDDOR)* (revised edition published 10/12/99).

RELATED DOCUMENTATION

- Records of any reported deaths, dangerous occurrences, injuries or diseases (which must be kept for at least 3 years).
- Reports, notification and any other documentation sent to the enforcing authority.
- Notification of any gas incidents.

2.2 Emergency Services Notifications

LEGISLATION

- Dangerous Substances (Notification and Marking of Sites) Regulations 1990 (SI 1990 No. 304).

KEY DATES

- In force from 1/9/90.

PURPOSE

- Ensure that emergency services and regulatory authorities have information on dangerous substances on site.
- Ensure that fire authorities entering the site in the event of an emergency are adequately warned of presence, location and nature of dangerous substances by means of appropriate marking of the site.
- Minimise health, safety and environmental consequences of accidents involving sites at which dangerous substances are stored.

APPROACH

- Prior written notice required to local fire authority and HSE when 25 tonnes, or more, of dangerous substances are to be present on site, and also any change or cessation to be reported to the regulatory authority.
- Appropriate marking of the site and warning signs giving information on the hazards present must be displayed at access to site and within site so as to give adequate warning to fire department and any other emergency services entering the site in the event of an emergency.

REGULATORY AUTHORITY

- Fire Authority.
- HSE.

COVERAGE

- Regulations apply to any site where there is a quantity (in total) of 25 tonnes, or more, of dangerous substances that fall within the definition of dangerous goods in Carriage of Dangerous Goods (Classification, Packaging and Labelling) and Use of Transportable Pressure Receptacles Regulations 1996.
- For the purpose of these Regulations, dangerous substances do not include substances which are buried as waste, radioactive substances, substances in certain types of aerosol dispensers or Class I explosives.
- There are certain exemptions from the notification requirements for particular sites including sites which have a current licence for keeping of substances under the Petroleum (Consolidation) Act 1928.

COMPLIANCE REQUIREMENTS

- Must notify regulatory authority prior to 25 tonnes, or more, of dangerous substances being present on site. This notification must include the following details:
  - The notifier's name and address.
  - The full postal address of the site.
  - A general description of the nature of business at the site.
  - A list of classification of the substances in question.
  - The date when the threshold will be passed.
- Must notify any reduction below the 25 tonnes threshold or cessation to regulatory authorities.
- Must display warning signs, (consisting of a yellow triangle with a black border and a black exclamation mark), at access points to the site, so as to give adequate warning to fire department and any other emergency services entering the site in the event of an emergency.
- An inspector may require the actual location of the dangerous substance on site be marked with more specific signs giving the nature of the hazards.
- The signs to be displayed should conform with Health and Safety (Safety Signs and Signals) Regulations 1996. Note: the inspector may instruct as to the location of signs.
- Signs must be kept clean and free from obstruction.

DISCLOSURE

- Notification to authorities and any amendments.

COMPLEMENTARY LEGISLATION

- Petroleum (Consolidation) Act 1928.
- Control of Major Accident Hazards Regulations 1999.
- Carriage of Dangerous Goods (Classification, Packaging, and Labelling) and Use of Transportable Pressure Receptacles Regulations 1996.
- Health and Safety (Safety Signs and Signals) Regulations 1996.

RELATED EC LEGISLATION

COMMENTS

- HSE can grant written exemption provided it is satisfied that the health and safety of persons likely to be affected will not be prejudiced.

RELATED DOCUMENTATION

- Notification.
- Notification of any change or cessation.
- Site map demonstrating location of signs.
- Dangerous substances inventory of chemicals on site.

2.3 Enforcement

LEGISLATION

- Environment and Safety Information Act 1988.

KEY DATES

- In force from 1/4/89.

PURPOSE

- Establish public registers of certain notices served concerning health, safety and environmental protection.

APPROACH

- Various public authorities with duties under health, safety and environmental legislation are required to maintain registers of certain notices served under such legislation. The notices covered include:
  - Improvement and prohibition notices under the Health and Safety at Work *etc.* Act 1974.
  - Prohibition notices under the Fire Precautions Act 1971.
- Provide notification system for persons affected by notice to have information excluded on grounds that it would disclose information about a trade secret or a secret manufacturing process. Person on whom a notice is served can, within 14 days, give written notification to the authority that entry on register would disclose such secrets. Authority will draft entry which, in their opinion, will not disclose the secret. A right of appeal to appropriate Minister against draft entry exists.

REGULATORY AUTHORITY

- Enforcing authority (HSE or local authorities) under Health and Safety at Work *etc.* Act 1974.
- Fire authority under Fire Precautions Act 1971.

COVERAGE

- Improvement and prohibition notices under Health and Safety at Work *etc.* Act 1974.
- Prohibition notice under Fire Precautions Act 1971.

COMPLIANCE REQUIREMENTS

DISCLOSURE

- Register and index are open to inspection by the public free of charge at all reasonable hours.
- Copies of entries are available to the public for a reasonable charge.

COMPLEMENTARY LEGISLATION

- Health and Safety at Work *etc.* Act 1974.
- Fire Precautions Act 1971.

RELATED EC LEGISLATION

COMMENTS

- Note that the Act does not cover improvement notices issued under the Fire Precautions Act 1971.

RELATED DOCUMENTATION

- Any improvement and prohibition notices under Health and Safety at Work *etc.* Act 1974.
- Any prohibition notice under Fire Precautions Act 1971.
- Any notification to authority that entry on register would disclose trade secrets.

2.4 Fire Certificates

LEGISLATION

- Fire Precautions Act 1971 as amended by the Fire Safety and Safety of Places of Sport Act 1987.

KEY DATES

- Most sections in force from 1/1/77.
- Amendment in force from 1/1/88.

PURPOSE

- To protect persons from fire risks at specified premises with a higher hazard associated with them.

APPROACH

- Premises meeting certain specified criteria must have a valid fire certificate.
- Creates certain specific criminal offences relating to fire matters, *e.g.* failure to maintain equipment, contravening the requirements of a fire certificate *etc.*

REGULATORY AUTHORITY

- Fire Brigades.
- HSE for *special premises*.

COVERAGE

- Premises defined as requiring a fire certificate are:
  - Certain hotels or boarding houses.
  - Workplaces which:
    - Employ more than 20 people at any one time, or
    - Employ more than 10 people at any one time other than on the ground floor, or

- Store or use explosive or flammable substances or ionising radiation – called *special premises* detailed in the Fire Certificate (Special Premises) Regulations 1976.

COMPLIANCE REQUIREMENTS

- Owner/occupier of premises must apply for a fire certificate to either the fire authority or the HSE and include certain specified information. This certificate must be kept on the premises when it is obtained.
- The owner/occupier must adhere to precautions contained in the fire certificate including details of means of escape, means to ensure efficacy of the fire escape, fire fighting equipment, fire alarm and training.

DISCLOSURE

COMPLEMENTARY LEGISLATION

- Fire Certificate (Special Premises) Regulations 1976.
- Fire Precautions (Workplace) Regulations 1997.
- Radiation (Emergency Preparedness and Public Information) Regulations 2001.

RELATED EC LEGISLATION

COMMENTS

- The fire authority can issue improvement notices (specifying matters to be rectified in a certain time) or prohibition notices (specifying a ban on the use of the premises until the fire authority is satisfied that the matters giving rise to a serious risk have been rectified).
- Radiation (Emergency Preparedness and Public Information) Regulations 2001 added premises covered by these 2001 Regulations to the list of special premises. These are premises involving work with ionising radiation.

RELATED DOCUMENTATION

- Fire certificate and application.
- Any improvement or prohibition notices.

2.5 Hazardous Substances Notification

LEGISLATION

- Notification of Installations Handling Hazardous Substances Regulations 1982 (SI 1982 No. 1357).

KEY DATES

- In force from 1/1/83.

PURPOSE

- To identify to HSE sites which due to the presence of hazardous substances on site present a potential risk to the safety of employees or the public.

APPROACH

- Notification to HSE required prior to commencing any proposed activity at a site or pipeline involving a notifiable quantity of a hazardous substance.

REGULATORY AUTHORITY

- HSE.

COVERAGE

- Hazardous substances are defined as named listed substances or by a generic hazard classification of substances *e.g.* toxic – both list notifiable quantities in the Regulations. The lowest notifiable quantity for a specific substance is 2 tonnes and for a generic class 15 tonnes.
- Regulations apply to any site, certain off-site pipelines and structures on inland waters.

COMPLIANCE REQUIREMENTS

- Must notify HSE at least 3 months prior to commencing any activity on site (or in certain pipelines or inland waters) involving a notifiable quantity of a hazardous substance with the following information:
  - The name and address of the applicant and the local planning authority.
  - The full postal address and Ordnance Survey grid reference of the site.
  - A general description of the proposed activity and its commencement date.
  - The minimum and maximum quantities of the relevant named substances.
- Must notify HSE of any change in notified activity.
- Must re-notify HSE if quantity of substance increases to three times or more than that already notified.

DISCLOSURE

- Notification to HSE.

COMPLEMENTARY LEGISLATION

- Control of Major Accident Hazards Regulations 1999 (COMAH).
- Dangerous Substances (Notification and Marking of Sites) Regulations 1990.

RELATED EC LEGISLATION

COMMENTS

- HSE may grant written exemptions for persons or activities if satisfied that this does not prejudice health and safety of persons affected.
- Licensed waste disposal sites are not covered by the Regulations.

RELATED DOCUMENTATION

- Any notification(s), re-notification(s) to HSE.
- Site chemical inventory of hazardous substances.

2.6 Notifications of Presence of Hazardous Substances/Planning

LEGISLATION

- Planning (Hazardous Substances) Act 1990.
- Planning (Hazardous Substances) Regulations 1992 (SI 1992 No. 656).
- Planning (Control of Major Accident Hazards) Regulations 1999 (SI 1999 No. 981).

KEY DATES

- 1990 and 1992 legislation in force from 1/6/92.
- 1999 Regulations in force from 20/4/99.

PURPOSE

- Regulate presence of hazardous substances on land to ensure only held in appropriate areas of community and to ensure their safe storage.
- To ensure that the objectives of both preventing major accidents and limiting their consequences are taken into account in land-use planning policies, and that these objectives are pursued through adequate controls and further consultation procedures.

APPROACH

- Under the Planning (Hazardous Substances) Act 1990, storage on site of listed hazardous substances in amounts equal to or greater than controlled quantities requires a hazardous substances consent which must be held and complied with.
- The Planning (Hazardous Substances) Regulations 1992 as amended by 1999 Regulations define hazardous substance and the procedures for obtaining the consent and bring the Act into force.

REGULATORY AUTHORITY

- HSA – Hazardous Substances Authority (usually the local planning authority) must consult with specified bodies, *e.g.* environment agency before granting consent.

COVERAGE

- Hazardous substances are defined to only include raw materials, products, by-products, residues or intermediates.
- Substances and preparations listed in the Regulations (when present in specified controlled quantities) either by name or category of hazard. Also includes substances used in industrial chemical processes if loss of control of the process may generate the listed substances and preparations.
- The lowest controlled quantities for specified substances are 0.001 tonnes for polychlorodibenzofurans and specified carcinogens, *e.g.* chloromethyl methyl ether, and for generic category, 5 tonnes for very toxic substances and preparations.
- Exemptions include explosives and radioactive waste as these are subject to separate legal requirements.

COMPLIANCE REQUIREMENTS

- Must apply to HSA for hazardous substances consent prior to the presence of a specified controlled quantity of a hazardous substance on, over or under land which forms part of a single establishment.
- Must apply for continuation of consent after a change in control of part of land.
- Fee payable to HSA for both above applications.
- Must advertise application in a local newspaper and post notice at site, both within specified times.
- Consent may be revoked/modified by HSA (see Comments).
- Conditions attached to consent relating to storage of substances must be complied with.

DISCLOSURE

- Advertise application in local newspaper and copy of application to be available for inspection at site.
- Public register of consents to be kept by HSA.

COMPLEMENTARY LEGISLATION

- Town and Country Planning Act 1990.
- Notification of Installations Handling Hazardous Substances Regulations 1982.
- Control of Major Accident Hazards Regulations 1999.

RELATED EC LEGISLATION

- EC Directive 96/82 on the control of major accident hazards involving dangerous substances.

COMMENTS

- Compensation payable in certain cases where consent revoked/modified by HSA.
- The list of hazardous substances includes those specifically listed in the Control of Major Accident Hazards Regulations 1999.

RELATED DOCUMENTATION

- Application for authorisation and related correspondence.
- Notice of application.
- Consent.
- Application for continuation of consent.
- Any modification/revocation.
- Site chemical inventory.

2.7 Petroleum

LEGISLATION

- Petroleum (Consolidation) Act 1928.

KEY DATES

- In force from 3/8/28.

PURPOSE

- To ensure local authority is aware of exact location of premises keeping petroleum-spirit and that it is kept safely.

APPROACH

- A licence is required to keep petroleum-spirit.
- Certain labelling requirements for vessels containing petroleum-spirit are specified.

REGULATORY AUTHORITY

- Local authority.

COVERAGE

- Petroleum includes crude petroleum, oil made from petroleum or from coal, shale, peat or other bituminous substances, and other products of petroleum.
- Petroleum-spirit means such petroleum which has a flash point of less than 21 degrees Celsius when tested in accordance with the Chemicals (Hazard Information and Packaging for Supply) Regulations 2002.

COMPLIANCE REQUIREMENTS

- No occupier of premises may keep petroleum-spirit unless a petroleum-spirit licence has been obtained, is in force and the terms of the licence are complied with.

- A licence is not required for petroleum-spirit kept for private use or for sale if it is in containers of not more than 570 mililitres. and the total amount does not exceed 15 litres.
- The local authority may impose licence conditions relating to storage, facilities for testing the petroleum-spirit and general safe keeping of petroleum-spirit.
- Where petroleum-spirit is kept, transported or offered for sale a label stating "Petroleum-spirit" and "Highly Inflammable" shall be attached to, or if impractical displayed near, the vessel containing the petroleum-spirit. Also the name and address of the owner, sender or vendor shall be stated, whichever is appropriate.

DISCLOSURE

COMPLEMENTARY LEGISLATION

- Health and Safety at Work *etc.* Act 1974.
- Chemicals (Hazard Information and Packaging for Supply) Regulations 2002 (CHIP 3).

RELATED EC LEGISLATION

COMMENTS

- Any appeal relating to the licence or conditions is to the Secretary of State under the Health and Safety at Work *etc.* Act 1974.
- The Act gives the local authority wide powers including the right to take samples of petroleum-spirit and have them tested, and to search for and seize petroleum-spirit after obtaining a court warrant.

RELATED DOCUMENTATION

- Licence application documentation.
- Testing results.
- Petroleum-spirit licence and conditions.
- Any petroleum inventory.

2.8 Prevention of Major Accidents

LEGISLATION

- Control of Major Accident Hazards Regulations 1999 (SI 1999 No. 743) (COMAH).

KEY DATES

- In force from 1/4/99.

PURPOSE

- Prevent and limit effects of accidents arising from activities involving dangerous substances.
- Implement EC Directive 96/82 except Article 12 relating to planning which is implemented by other regulations.

APPROACH

- Imposes general requirements on operators of establishments where dangerous substances in specified quantities are present.
- Imposes more stringent requirements for establishments where specified high quantities of dangerous substances are present.

REGULATORY AUTHORITY

- HSE.
- Environment Agency.
- Local authority.
- Emergency services.

COVERAGE

- Operators meaning a person who is in control of an establishment or installation where dangerous substances are present.
- Dangerous substances (include a substance, mixture or preparation) are listed by name in the Regulations *e.g.* ethylene oxide or by hazard category, *e.g.* toxic.

- The Regulations establish two tiers of requirements. The lower tier requirements apply to dangerous substances in specified quantities listed in the Regulations, *e.g.* 5 tonnes or more of ethylene oxide present and the higher tier specifies higher quantities, *e.g.* 50 tonnes or more of ethylene oxide.
- The lowest specified quantity is 0.001 tonnes for named substances including dioxin and certain carcinogens.
- Certain establishments are exempt, *e.g.* Ministry of Defence installations and waste landfill sites.

COMPLIANCE REQUIREMENTS

- Operators of lower and higher tier establishments must:
  - Take all necessary measures to prevent major accidents and limit consequences to persons and the environment.
  - Prepare and keep a major accident prevention policy document which demonstrates establishment of a safety management system taking into account specified principles, *e.g.* monitoring performance.
  - Within a reasonable time prior to start of construction (or for existing establishments by 3/2/00) send to the HSE and EA specified information *e.g.* quantity and physical form of dangerous substance present, name and address of operator and address of the establishment. Thereafter significant changes need to be notified. Information sent in a safety report need not be repeated.
  - Provide specified information to the HSE and/or EA if they request it.
  - Forthwith notify HSE and EA if a major accident has occurred.
- Operators of higher tier establishments must:
  - Prior to the start of construction (and prior to the start of operation) send to the HSE and EA a safety report with sufficient specified information *e.g.* demonstrate adequate safety and reliability incorporated into design. Construction must not start until the HSE and EA have sent their conclusions on their examination of the report.

- If the establishment is existing send to HSE and EA a similar safety report within specified time frames, *e.g.* if the establishment were covered by the previous legislation (CIMAH) by 3/2/00 and in any event by 3/2/01.
- Safety reports must be reviewed and revised as specified at least every 5 years.
- Prepare an on-site and off-site emergency plan after consultation with specified bodies *e.g.* the emergency services. These must be reviewed and tested at least every 3 years.
- Put emergency plans into effect if a major accident or uncontrolled event occurs.
- Provide after consultation with the local authority specified information to persons likely to be affected and to the public. This information to be reviewed at least every 3 years.

DISCLOSURE

- HSE or EA may request further information on any matter within the remit of these Regulations.
- Must inform persons working on site (including contractors) of major accident hazards and emergency plans.
- Information to the public concerning higher tier sites.

COMPLEMENTARY LEGISLATION

- Control of Industrial Major Accident Hazards 1994 (CIMAH) (repealed by COMAH but relevant for transitional provisions for sites previously covered by CIMAH).
- Chemicals (Hazard Information and Packaging for Supply) Regulations 2002 (CHIP 3).
- Planning (Control of Major Accident Hazards) Regulations 1999.

RELATED EC LEGISLATION

- EC Directive 96/82 on the control of major accident hazards involving dangerous substances.

COMMENTS

- Categories of substances are classified in accordance with CHIP 3.
- The Regulations specify functions of the HSE and EA and give them wide powers including prohibiting the operation of an establishment.
- Guidance on the interpretation of major accident to the environment for the purposes of the COMAH Regulations has been issued by the Department of the Environment.

RELATED DOCUMENTATION

- Major accident prevention policy document.
- Safety report(s).
- Any notification of major accidents.
- Information and training records.
- Information to persons likely to be affected and the public.
- On-site emergency plan.
- Off-site emergency plan.
- Any notification of changes.
- Any information requested by HSE or EA.
- Site chemical inventory.

2.9 Radiation – Emergency Provisions

LEGISLATION

- Radiation (Emergency Preparedness and Public Information) Regulations 2001 (SI 2001 No. 2975).

KEY DATES

- In force from 20/9/01.

PURPOSE

- Protect the health of workers and the general public from the dangers arising from ionising radiation.
- Implement Title IX, Section 1 (Intervention in cases of radiological emergency) of EC Directive 96/29/Euratom laying down basic safety standards for ionising radiation.

APPROACH

- Impose requirements on operators of premises where radioactive substances are present including risk assessments and emergency plans.
- Impose requirements on carriers transporting radioactive substances.

REGULATORY AUTHORITY

- HSE.
- Local authority.

COVERAGE

- Radioactive substances that contain one or more radionuclides whose activity cannot be disregarded for the purposes of radiation protection.
- Ionising radiation means the transfer of energy in the form of particles or electromagnetic waves of a wavelength of 100 nano-

metres, or less, or a frequency of 3×10^{15} hertz or more, capable of producing ions directly or indirectly.

- Includes self-employed as well as employed.

COMPLIANCE REQUIREMENTS

- Must where the work or transport (including through any public place) involves ionising radiation (as detailed in Regulation 3) make an assessment as to hazard identification and risk evaluation and, where the assessment reveals a radiation risk, to take all reasonably practicable steps to prevent a radiation accident or limit the consequences should an accident occur.
- Must send to the HSE a report of the above assessment containing specified matters at specified times (normally at least 12 months before the commencement of work with ionising radiation). HSE can ask for further particulars as they may reasonably require.
- Must make a further reassessment if any major changes to work with ionising radiation and in any event at least 3 years after first assessment.
- If assessment reveals a reasonably foreseeable radiation emergency arising then operator or carrier and, in the case of an operator, the local authority in whose area the premises are situated, must prepare emergency plans. These plans shall contain specified information including arrangements for emergency exposure and arrangements for off-site mitigatory action. These plans must be reviewed, revised and tested at intervals not exceeding 3 years.
- Must put emergency plan into action without delay if a radiation emergency occurs or an event occurs which could reasonably lead to such an emergency. Must then notify HSE without delay.
- Where the emergency plan provides for the possibility of any employee receiving an emergency exposure then such employees shall be identified. Specified arrangements shall be made for these employees including training, making equipment available to restrict exposure, dose assessments and medical surveillance.
- Must pay local authority a fee for emergency plan work if one is requested.

- Must supply specified information to the public where there is a possibility of a reasonably foreseeable radiation emergency or where there has been a radiation emergency.

DISCLOSURE

- Public information.
- Emergency plan.
- Accident report.

COMPLEMENTARY LEGISLATION

- Management of Health and Safety at Work Regulations 1999.

RELATED EC LEGISLATION

- EC Directive 96/29/Euratom laying down basic safety standards for the protection of the health of workers and the general public against the dangers arising from ionising radiation.

COMMENTS

- The requirements of these Regulations are without prejudice to the Management of Health and Safety at Work Regulations 1999 which provide for risk assessments to be undertaken for work activities.

RELATED DOCUMENTATION

- Accident report.
- Emergency plan.
- Public information.
- Site inventory of radioactive substances.
- Carrier/transport documentation.
- Data and information on radioactive substance.
- Personnel records of relevant employees.
- Training records.
- Medical surveillance records.

2.10 Radioactive Substances Control

LEGISLATION

- Radioactive Substances Act 1993.
- Radioactive Substances (Clocks and Watches) (England and Wales) Regulations 2001 (SI 2001 No. 4005).

KEY DATES

- In force from 27/8/93.
- 2001 Regulations in force from 14/1/02.

PURPOSE

- Regulate keeping and use of radioactive material and the accumulation and disposal of radioactive waste.

APPROACH

- Registration required for use or keeping of radioactive material on premises and of mobile radioactive apparatus.
- Authorisation required for disposal and accumulation of radioactive waste.

REGULATORY AUTHORITY

- Environment Agency (EA).
- Local authority.
- HSE.

COVERAGE

- Radioactive material is defined as anything which not being waste, is either a substance (or an article made wholly or partly from, or incorporating, such a substance):
  - Containing an element listed in Schedule I (in sufficient proportion),

- or that possesses radioactivity due to nuclear fission or other non-natural process.
- Radioactive waste is a substance which, if it was not a waste, would be radioactive material or a substance or article contaminated in the course of producing, keeping or using radioactive material.
- Exemption is provided for clocks and watches but where only tritium, promethium or radium is present and either the amount of radionuclides on premises does not exceed specified limits or the waste is from no more than five clocks or watches.
- Applies to any person using premises for the purposes of an undertaking by him.

COMPLIANCE REQUIREMENTS

- Must apply for registration of premises to EA to keep or use radioactive material.
- Must apply for registration to EA to keep, use, lend or let on hire any mobile radioactive apparatus.
- Must comply with conditions/limitations imposed by EA.
- Charges payable to EA.
- Must apply for authorisation to accumulate or dispose of radioactive waste.
- Regulatory authority can revoke, or vary, conditions in registration or authorisation.
- Local authority can take special precautions to ensure radioactive waste is disposed of in accordance with authorisation.

DISCLOSURE

- Copies of certificate of registration or authorisation must be posted on premises.
- Regulatory authority may require retention and production of site or disposal records.
- Various documents must be sent to local authority and made available to the public.

COMPLEMENTARY LEGISLATION

- Nuclear Installations Act 1985.
- Transfrontier Shipment of Radioactive Waste Regulations 1993.
- Environment Act 1995.

RELATED EC LEGISLATION

- EC Directive 96/29 Euratom – Basic Safety Standards.

COMMENTS

- Grounds of appeal exist against decisions of the EA in relation to registrations/authorisations, any attached conditions/limitations, or cancellations and revocations.
- 2001 Regulations restrict the clocks and watches exemption to comply with EC Directive 96/29.

RELATED DOCUMENTATION

- Application for registration of premises and related correspondence.
- Registration.
- Application for authorisation and related correspondence.
- Authorisation.
- Any revocation/variation notice served.
- Any correspondence with local authority relating to disposal of radioactive waste.
- Copies of certificate of registration or authorisation posted on premises.
- Site/disposal records.

3 SUBSTANCES

3.1 Asbestos – Licensing

LEGISLATION

- Asbestos (Licensing) Regulations 1983 (SI 1983 No. 1649).
- Asbestos (Licensing) (Amendment) Regulations 1998 (SI 1998 No. 3233).

KEY DATES

- In force from 1/8/84.
- (Amendment) Regulations 1998 in force from 1/2/99.

PURPOSE

- Ensure that asbestos removal projects and procedures, and other works involving asbestos are tightly controlled and monitored.

APPROACH

- Imposes general requirement that employer or self-employed person must obtain licence from HSE before undertaking work with asbestos insulation, asbestos coating or asbestos insulating board (so that work is usually carried out by specialised licensed contractors).
- Alternatively, where employer or self-employed person carrying out work with asbestos insulation, asbestos coating or asbestos insulating board at own premises, may notify enforcing authority specifying work to be carried out and comply with other training and information requirements.

REGULATORY AUTHORITY

- HSE.
- EHOs.

COVERAGE

- Asbestos insulation, asbestos coating, asbestos insulating board except asbestos cement.
- The 1998 Regulations amend the definitions in the 1983 Regulations of "work with asbestos insulation and asbestos coating" and work with "asbestos insulating board" so as to cover only the work which consists of the removal, repair or disturbance of such asbestos, and work which is ancillary to or supervising such work.

COMPLIANCE REQUIREMENTS

- Must obtain licence from HSE before undertaking work with asbestos insulation, asbestos coating or asbestos insulating board, and comply with the terms and conditions of that licence.
- Where employer or self-employed person carrying out work with asbestos insulation, asbestos coating or asbestos insulating board at own premises, and valid licence not held, must notify enforcing authority in writing at least 14 days in advance specifying work to be carried out. Must also provide adequate information to persons likely to be affected, adequate training and instruction for employees, and ensure that both persons likely to be affected and employees are exposed only to the lowest level of asbestos dust which is reasonably practicable.
- The 1998 Regulations extend the circumstances in which the HSE may revoke a licence. The HSE can now revoke a licence if "it considers it appropriate to do so".

DISCLOSURE

COMPLEMENTARY LEGISLATION

- Control of Asbestos at Work Regulations 2002.

RELATED EC LEGISLATION

COMMENTS

- (Amendment) Regulations 1998 extend the 1983 Regulations so as to prohibit an employer or self-employed person carrying out work with asbestos *insulating board* without a licence, require such work to be notified to the enforcing authority, and provide for exemption from such a prohibition and requirement to be granted by the HSE. The period for notification to the enforcing authority of work with asbestos insulation, asbestos coating or asbestos insulating board is also reduced from 28 to 14 days.
- HSE may grant exemptions for persons or products containing asbestos if satisfied that this will not prejudice health and safety of persons affected.

RELATED DOCUMENTATION

- Licence from HSE (or copy of any contractor's licence).
- Notification to enforcing authority.
- Information to persons likely to be affected.
- Employee training records.

3.2 Asbestos – Work

LEGISLATION

- Control of Asbestos at Work Regulations 2002 (SI 2002 No. 2675).

KEY DATES

- In force from 21/11/02 except Regulation 4 (non-domestic premises duties) in force from 21/5/04 and Regulation 20 (analysis standards) in force from 21/11/04.

PURPOSE

- Protect workers from the risks related to exposure to asbestos.
- Implement various EC Directives, in particular EC Directive 83/477 on the protection of workers from the risks related to exposure to asbestos at work.

APPROACH

- Imposes various duties on all employers to protect employees who may be exposed to asbestos at work and also to protect other persons who may be affected by the work activity.

REGULATORY AUTHORITY

- HSE.
- HSC.

COVERAGE

- Asbestos for the purposes of the Regulations refers to crocidolite, amosite, chrysotile, fibrous actinolite, fibrous anthophyllite, fibrous tremolite and any mixture containing any of those minerals.

- Application of Regulations now extended to all employees exposed to asbestos (whether or not they are working with asbestos or any product containing asbestos).

COMPLIANCE REQUIREMENTS

- Dutyholder (person with repair and/or maintenance responsibilities for non-domestic premises) must ensure that a suitable and sufficient assessment is carried out as to whether asbestos is or is liable to be present in the premises. Asbestos-containing materials within these premises must be properly managed, and information about the location and condition of the materials must be passed on to those likely to disturb them.
- Employers carrying out work with asbestos must identify type of asbestos involved, and assess the degree of exposure to asbestos and the steps to be taken to reduce that exposure. This risk assessment must consider specified matters including monitoring of exposure and the effect of control measures. It must be regularly reviewed and forthwith if certain conditions occur *e.g.* significant change in the work to which the risk assessment relates.
- Must draw up plan of work, detailing how the work is to be carried out and various requirements including the equipment to be used, where the work involves removal of asbestos from plant or buildings.
- Employers must keep a copy of their risk assessment and plan of work at the place to which they relate and while the work to which they relate is being carried out, and to ensure that, so far as is reasonably practicable, the work is carried out in accordance with the plan.
- Must notify work with asbestos to regulatory authority (unless work notified under other regulations). The period for advance notification to the enforcing authority is 14 days.
- Must ensure that adequate information and training is given to employees and other persons on premises where the work is carried out.
- Must prevent exposure of employees and other persons to asbestos or reduce it to lowest level reasonably practicable. If exposure

is not reduced to below specified control limits, employees must be provided with respiratory protective equipment.

- Employer who provides any control measure shall ensure, so far as is reasonably practicable, that it is properly used or applied.
- Employer shall ensure that any control measure is maintained in an efficient state, in efficient working order, in good repair and in clean condition.
- Employer must prepare procedures, provide information and establish warning systems to deal with an emergency in the workplace related to the use of asbestos in a work process or the removal or repair of asbestos containing materials.
- Employees (as well as employers) are under duty to make proper use of protective equipment.
- Must provide and keep clean protective clothing (as well as washing and changing facilities), prevent the spread of asbestos from the workplace and ensure that premises and plant involved in work with asbestos are kept clean.
- Must monitor exposure to asbestos by measurement of asbestos fibres in the air except under certain conditions *e.g.* exposure is not likely to exceed an action level. Employers carrying out air monitoring to comply with, or to use European Standard ISO 17025.
- Employers must comply with duties concerning health records and medical surveillance. If medical surveillance reveals an employee has an identifiable disease or adverse health effect as a result of exposure to asbestos then employee must be informed, risk assessment reviewed, consider assigning employee to alternative work and review the health and safety of other employees likely to be affected.
- Must regulate the storage and distribution of raw asbestos.
- Labelling of containers containing raw asbestos and asbestos waste shall be in accordance with the Chemicals (Hazard Information and Packaging for Supply) Regulations 2002 and use the label specified in the 1987 Regulations.

DISCLOSURE

- Notification to enforcing authority.

COMPLEMENTARY LEGISLATION

- Asbestos (Licensing) Regulations 1983.
- Chemicals (Hazard Information and Packaging for Supply) Regulations 2002.

RELATED EC LEGISLATION

- EC Directive 76/769 on marketing and use of certain dangerous substances and preparations as amended by EC Directives 83/478 and 99/43.
- EC Directive 83/477 on the protection of workers from the risks related to exposure to asbestos at work as amended by EC Directive 91/382.
- EC Directive 90/394 on the protection of workers from the risks related to carcinogens at work.
- EC Directive 98/24 on the protection of the health and safety of workers from risks related to chemical agents at work.

COMMENTS

- Reference should be made to the ACOP L27 *Control of Asbestos at Work*.
- These Regulations replaced the Control of Asbestos at Work Regulations 1987 and introduced the requirements concerning non-domestic premises and emergency arrangements.

RELATED DOCUMENTATION

- Assessments of exposure to asbestos.
- Plan of work where the work involves removal of asbestos from plant or buildings, detailing how the work is to be carried out and various requirements including the equipment to be used.
- Notification to enforcing authority (unless work notified under other Regulations).
- Information and training records for employees and other persons on premises.
- Monitoring data relating to employees' exposure to asbestos.

- Records relating to provision of, and maintenance of respiratory protective equipment, clean protective clothing (as well as washing and changing facilities).
- Medical surveillance records (maintained by employment medical adviser or appointed doctor) for employees who are liable to be significantly exposed to asbestos.
- Records relating to the storage, distribution and labelling of raw asbestos.

3.3 Biocidal Products

LEGISLATION

- Biocidal Products Regulations 2001(SI 2001 No. 880).

KEY DATES

- In force from 6/4/01.

PURPOSE

- Ensure sufficient information is known about biocidal products to enable them to be marketed and used safely.
- Implement EC Directive 98/8 concerning the placing of biocidal products on the market.

APPROACH

- Placing on the market or use of a biocidal product or active ingredient is banned unless authorised or included in a European approved list.

REGULATORY AUTHORITY

- Secretary of State and the Minister for the Environment, Food and Rural Affairs acting jointly.

COVERAGE

- Biocidal product means an active substance or preparation intended to destroy, deter, render harmless, prevent the action of, or otherwise exert a controlling effect on, any harmful organism by chemical or biological means.
- Numerous exemptions are listed in Schedule 2 and are biocidal products covered by other legislation *e.g.* medicinal products, food flavourings, medical devices, veterinary medicinal products, homoeopathic products, feeding stuffs.

- Transport by rail, road, inland waterway, sea or air of biocidal products is exempt.
- New active substance means an active substance which was not on the market in the European Community before 14th May 2000 other than for research purposes.
- Placing on the market means any supply within Great Britain including importation and any subsequent storage.

COMPLIANCE REQUIREMENTS

- Must not place on the market a new active substance for use in a biocidal product unless an application to the competent authority (Ministers) has been made for inclusion of that new active substance in Annex I, IA or IB of EC Directive 98/8 and the Ministers have agreed the application. The application must be accompanied by dossiers containing specified information which the Ministers must evaluate and recommend whether or not the active substance should be included in the Directive's list.
- Must not place on the market or use a biocidal product unless it has been authorised or registered. A registration is only applicable for low-risk biocidal products as defined. To obtain an authorisation or registration an application must be submitted and approved with specified information. The authorisation or registration may be approved subject to conditions.
- Must provide information to the National Poisons Information Service if first person placing biocidal product on the market.
- Must comply with specified packaging, labelling and advertising requirements.
- Must pay fees.

DISCLOSURE

- Dossier for new active substance.
- Registration or authorisation application.
- Information to National Poisons Information Service.

COMPLEMENTARY LEGISLATION

RELATED EC LEGISLATION

- EC Directive 98/8 concerning the placing of biocidal products on the market.

COMMENTS

- Ministers may grant an emergency authorisation where it appears necessary because of unforeseen danger.
- Ministers may prohibit or restrict the sale or use of a biocidal product if they consider it constitutes an unacceptable risk to human or animal health or to the environment.
- Information provided to Ministers can be kept confidential if the applicant can justify that such information might harm his industrial or commercial position.
- A right of appeal against the Ministers decisions is provided for.

RELATED DOCUMENTATION

- New active substance research data.
- Dossier application and related correspondence.
- Authorisation or registration information and related correspondence.
- Production data.
- Labelling, packaging and advertisement data.

3.4 Classification, Packaging and Labelling

LEGISLATION

- Chemicals (Hazard Information and Packaging for Supply) Regulations 2002 (SI 2002 No. 1689) (CHIP 3).

KEY DATES

- In force from 24/7/02.

PURPOSE

- Ensure the adequate classification, packaging and labelling of dangerous substances for supply, in line with EU and internationally agreed standards, enabling them to be used safely.

APPROACH

- CHIP applies if the substance or preparation is dangerous for supply and then it must be classified with the hazards that the chemical(s) represent.
- Suppliers must package and label the substance or preparation in a prescribed manner.
- Suppliers must provide safety data sheets to recipients of the substance or preparation to inform them of the chemical's hazards, safety precautions *etc*.

REGULATORY AUTHORITY

- HSE.

COVERAGE

- Regulations apply to supply of any dangerous substance or preparation, meaning substances listed in Part I of the Approved Supply List (ASL) (see Comments) or substances/preparations within categories of danger in Schedule 1 of the Regulations. (Substances are chemical elements/compounds in a natural or produced state, while preparations are mixtures or solutions of two or more substances). Numerous exceptions from the Regulations include certain medicinal products, cosmetics, foodstuffs, pesticides, wastes, animal feedstuffs and munitions.

- "Supply" means making a substance or preparation available to another person and includes importation of the substance or preparation into Great Britain.

COMPLIANCE REQUIREMENTS

- Suppliers of all substances or preparations dangerous for supply must classify the hazards that the chemicals represent before supply by:
  - If substance is listed in the ASL classifying according to the entry in this list (see Comments).
  - If substance has been notified in accordance with the Notification of New Substances Regulations 1993 classifying in conformity with that notification.
  - If the substance is not in the ASL or is a preparation then classify by taking into account all relevant and accessible data. For substance the Approved Classification and Labelling Guide (ACLG) explains how to classify and gives criteria for classification. For preparations classify by either using a calculation method, or for certain hazards use the criteria in ACLG.
- Suppliers must label a dangerous chemical if it is supplied in a package. The Regulations specify what has to go on the label, including risk and safety phrases, and how packages should be labelled including the size of the label.
- Suppliers must provide a safety data sheet to recipients, who are going to use the chemical for their work, to inform them of the chemical's hazards. These sheets must contain information under 16 prescribed headings (including data on transport and first aid) which will enable the recipient to take the necessary measures to protect health, safety and the environment.
- Suppliers must package the chemicals in a prescribed and suitable manner which is sufficient to prevent them escaping.

- Any advert for a dangerous substance or preparation must mention the type of hazard indicated on the label.
- Suppliers must retain the data used for the classification and preparation of the safety data sheet for at least 3 years after the substance or preparation was last supplied. Such information must be provided to HSE within 28 days if they request it.
- Certain substances require additional labelling *i.e.* "Restricted to professional users". These substances are named carcinogens, mutagens and certain substances toxic for reproduction.
- Certain specified substances, either by hazard or chemical concentration, require child resistant closures and/or tactile warning devices if supplied to the general public.

DISCLOSURE

- Information supplied to HSE on request.

COMPLEMENTARY LEGISLATION

- Carriage of Dangerous Goods (Classification, Packaging and Labelling) and Use of Transportable Pressure Receptacles Regulations 1996.
- Notification of New Substances Regulations 1993 (NONS).
- Carriage of Dangerous Goods (Amendment) Regulations 1999.
- Control of Substances Hazardous to Health Regulations 2002 (COSHH).
- Chemicals (Hazard Information and Packaging for Supply) Regulations 1994 (CHIP 2), Chemicals (Hazard Information and Packaging for Supply) (Amendment) Regulations 1996 (CHIP 96) and (Amendment) Regulations 1997 (CHIP 97), (Amendment) Regulations 1998 (CHIP 98), (Amendment) Regulations 1999 (CHIP 99), (Amendment) (No. 2) Regulations 1999 (CHIP 99/2) and (Amendment) Regulations 2000.
- Biocidal Products Regulations 2001.

RELATED EC LEGISLATION

- EC Directive 67/548 on the Classification, Packaging and Labelling of Dangerous Substances as amended for the seventh time by EC Directive 92/32 (the Substances Directive) and numerous Adaptations to Technical Progress (ATP).
- EC Directive 88/379 on the Classification, Packaging and Labelling of Dangerous Preparations.
- EC Directive 91/155 on Safety Data Sheets and its Amendments (93/112 and 2001/58).
- EC Directive 99/43 amending for the 17th time EC Directive 76/769 on the approximation of the laws, regulations and administrative provisions of the Member States relating to restrictions on the marketing and use of certain dangerous substances and preparations.
- EC Directive 99/45 on the Classification, Packaging and Labelling of Dangerous Preparations (the preparations Directive) and the first ATP EC Directive 2001/60.

COMMENTS

- The Regulations are supplemented by an Approved Supply List (ASL) ("*Information Approved for the Classification and Labelling of Substances and Preparations Dangerous for Supply*") (7th Edition) and the *Approved Classification and Labelling Guide* (5th Edition). These contain classification and labelling requirements agreed at EC level for specific substances. The ASL is based on ATP Directives.
- HSE may grant exemptions by a written certificate for persons, substances or preparations if satisfied that this will not prejudice health and safety of persons affected.
- CHIP 3 repeals CHIP 2 and all its additions and amendments. One major change is CHIP 3 now includes biocidal products. A transitional period exists for biocidal products within the meaning of the Biocidal Products Regulations 2001. Until 30th July 2004 it shall be sufficient to comply with CHIP 3 if the requirements of CHIP 2 are complied with immediately before CHIP 3 came into force. Hence CHIP 2 and associated regulations are listed under Complementary Legislation.

RELATED DOCUMENTATION

- Data used for classification.
- Safety data sheets.
- Packaging and labelling records.
- Any information supplied to HSE.

3.5 Dangerous Goods Carriage

LEGISLATION

- Carriage of Dangerous Goods (Classification, Packaging and Labelling) and Use of Transportable Pressure Receptacles Regulations 1996 (SI 1996 No. 2092).

KEY DATES

- In force from 1/9/96 except regulation 21(9) which is in force from 1/1/99.

PURPOSE

- Ensure that dangerous goods to be carried by road and rail are adequately classified, packaged and labelled in line with EU and internationally agreed standards.

APPROACH

- Require the consignor of dangerous substances that are to be carried by road and rail to ensure that the substances are adequately classified, packaged and labelled in a prescribed manner.
- Impose further requirements relating to the design, construction, modification, repair, approval, certification, examination, filling and marking of transportable pressure receptacles.

REGULATORY AUTHORITY

- HSE.
- HSC.

COVERAGE

- Regulations apply to dangerous goods meaning any explosive, radioactive material, goods listed in the Approved Carriage List (see Comments), or other goods that have one or more hazardous properties, *e.g.* corrosive substance, oxidising agent, as listed in the Regulations. The carriage of certain environmentally hazardous substances are included.

- There are numerous exceptions including the carriage of explosives and certain radioactive materials, and certain international transport operations.
- Transportable pressure receptacles (other than aerosols) used for the carriage of gas, with capacity above certain specified volumes.

COMPLIANCE REQUIREMENTS

- Must classify the hazards that the dangerous goods represent prior to their carriage, either by drawing on entries in the Approved Carriage List (see Comments) or by classifying the substance on the basis of its most hazardous property in accordance with Schedule 1 of the Regulations.
- Must package the dangerous goods in packages which are suitable for the purpose and comply with specified requirements, including that the packaging is sufficient to prevent the goods from escaping. The consignor must ensure that the packagings are used in accordance with various requirements, including the appropriate Approved Method (see Comments).
- Must label, classify and mark the packages in a prescribed manner. The label must contain various details including the designation of the goods, UN Serial Number, danger sign and subsidiary hazard sign.
- Comply with various requirements (including Approved Requirements) relating to the design, construction, modification, repair, approval, certification, examination, filling and marking of transportable pressure receptacles.

DISCLOSURE

COMPLEMENTARY LEGISLATION

- Carriage of Dangerous Goods by Rail Regulations 1996.
- Carriage of Dangerous Goods by Road Regulations 1996.
- Carriage of Dangerous Goods (Amendment) Regulations 1998 and 1999.
- Chemicals (Hazard Information and Packaging for Supply) Regulations 2002.

RELATED EC LEGISLATION

- EC Directive 94/55 on the approximation of the laws of Member States with regard to the transport of dangerous goods by road.
- EC Directive 96/49 on the approximation of the laws of Member States with regard to the transport of dangerous goods by rail.

COMMENTS

- Approved Carriage List refers to *Information Approved for the Classification, Packaging and Labelling of Dangerous Goods for Carriage by Road and Rail*. The appropriate Approved Method refers to method designated in *Approved Requirements and Test Methods for the Classification and Packaging of Dangerous Goods for Carriage*. Approved Requirements refers to *Approved Requirements for Transportable Pressure Receptacles*.
- HSE may grant written exemption certificates for persons or dangerous goods if satisfied that this will not prejudice health and safety of persons affected.

RELATED DOCUMENTATION

- Data used for classification.
- Packaging and labelling records.
- Pressure receptacle documentation including certificate, specification and maintenance record.

3.6 Hazardous Substances

LEGISLATION

- Control of Substances Hazardous to Health Regulations 2002 (SI 2002 No. 2677) (COSHH).

KEY DATES

- In force from 21/11/02.

PURPOSE

- Protect employees and other persons likely to be affected against risks to their health resulting from exposure to substances hazardous to health.

APPROACH

- Require all employers to assess and control the risks to health of all work activities which involve certain specified substances hazardous to health.
- Places duties on employers to prevent, and if that is not reasonably practicable control, the exposure of their employees – and anyone else likely to be affected – to substances hazardous to health by ensuring compliance with exposure limits, engineering controls, use of personal protective equipment (PPE) *etc*.
- Prohibit specified uses and import into UK, from non-EC countries, of certain substances.
- Employees are also placed under certain duties *e.g.* to use protective equipment properly.

REGULATORY AUTHORITY

- HSE.
- HM Customs and Excise regarding import bans.

COVERAGE

- Substance means a natural or artificial substance whether in solid or liquid form or in the form of a gas or vapour (including microorganisms).
- Preparation means a mixture or solution of two or more substances.
- Substances hazardous to health (which includes preparations) defined as:
  - Substances listed in the Approved Supply List (ASL) as dangerous for supply (*i.e.* very toxic, toxic, harmful, corrosive or irritant) under the Chemicals Hazard Information and Packaging for Supply Regulations 2002 (CHIP 3). This includes substances that have chronic or delayed effects *e.g.* carcinogens.
  - A substance for which the HSC has approved a maximum exposure limit (MEL) or an occupational exposure standard (OES). These substances are published in *EH40, Occupational Exposure Limits*.
  - A biological agent.
  - Dusts when present at substantial concentrations in the air. Detailed guidance is given in COSHH.
  - Any other substance that creates a risk to health to any person which is comparable with the hazards created by the above substances.
- Exemptions include where other Regulations apply *e.g.* asbestos and lead (see under Complementary Legislation), certain radioactive and explosive substances and where risk is due to the substance being administered in the course of medical treatment.

COMPLIANCE REQUIREMENTS

- Employer must not use certain specified substances for listed uses *e.g.* hydrogen cyanide for use in fumigation except if released from an inert material in which hydrogen cyanide is absorbed.
- Importation of named substances is prohibited except from another member state. These include 2-naphthylamine and matches made with white phosphorus.

- Assessment of health risks and steps needed to meet COSHH must be carried out prior to employees, or any other person, being exposed to substances hazardous to health and these must be reviewed regularly and forthwith in certain instances. Risk assessment shall be recorded where there are five or more employees.
- Employers must prevent or control exposure to substances hazardous to health. Prevention is achieved by substitution of the substance hazardous to health. Control can be achieved either by engineering controls, operating procedures or a combination of both. There is a hierarchy of measures – prevention of exposure must where reasonably practicable take precedence over any method that controls or limits exposure. PPE is only to be used as a last resort. The measures the employer must take to prevent or adequately control the exposure are detailed including specific requirements for transport, carcinogens and biological agents.
- Employer shall ensure that any control measure is maintained in an efficient state, in efficient working order, in good repair and in a clean condition. Any engineering controls must undergo a thorough examination and testing at appropriate intervals. Circumstances are detailed where health surveillance is appropriate.
- Employers must ensure that the contents of containers and pipes for substances hazardous to health used at work are clearly identifiable.
- Employer must prepare procedures, provide information and establish warning systems to deal with an emergency in the workplace related to the presence of a substance hazardous to health.
- Employer who provides any control measures, *e.g.* PPE, shall take all reasonable steps to ensure that it is properly used or applied.
- Employer to monitor employee exposure to substances hazardous to health and provide appropriate health surveillance.
- Personal health surveillance records to be kept for at least 40 years.

- Employer to provide employees who may be exposed to substances hazardous to health with such information, instruction and training as is suitable and sufficient for him to know risk to health and precautions to be taken.
- Employees are also placed under certain duties *e.g.* to use protective equipment properly.
- Special provisions relating to biological agents are extensively detailed in COSHH, *e.g.* lists measures to be applied where exposure cannot be prevented including displaying a biohazard sign.
- Special control measures are specified for health and veterinary care facilities, laboratories, animal rooms and industrial processes using biological agents.

DISCLOSURE

COMPLEMENTARY LEGISLATION

- Control of Asbestos at Work Regulations 2002.
- Control of Lead at Work Regulations 2002.
- Chemicals (Hazard Information and Packaging for Supply) Regulations 2002.

RELATED EC LEGISLATION

- EC Directive 80/1107 on the protection of workers from the risks related to exposure to chemical, physical and biological agents at work and associated EC Directives 91/322 and 96/94 establishing indicative limit values and EC Directive 88/364 banning certain specified agents and/or certain work activities.
- EC Directive 78/610 concerning vinyl chloride and EC Directive 89/677 concerning benzene.
- EC Directive 90/394 on the protection of workers from the risks related to exposure to carcinogens at work.
- EC Directive 90/679 on protection of workers from risks related to exposure to biological agents at work.

- EC Directives 96/55 and 99/43 adapting to technical progress, EC Directive 76/769 on marketing and use.
- EC Directive 98/24 on the protection of the health and safety of workers from risks relating to chemical agents at work.
- EC Directive 2000/54 on the protection of workers from risks related to exposure to biological agents at work.

COMMENTS

- These Regulations repealed COSHH 1999 and introduced the requirements concerning emergency arrangements. Any record keeping requirements under COSHH 1999 are to continue as if these Regulations had not been passed.
- A substance hazardous to health includes chemical substances, mixtures, allergens, fumigants, biological agents if they fulfil one of the qualifying criteria listed above.
- Reference could be made to *A Step-by-step Guide to COSHH Assessment (HS(G)97)* (reprinted with amendments in November 1999).

RELATED DOCUMENTATION

- COSHH assessments.
- Monitoring information.
- Information and training records.
- Health surveillance and employee records.
- Emergency procedures, information *etc*.
- Engineering controls testing and examination records.
- PPE cleaning, inspection records.

3.7 Highly Flammable Liquids and Liquefied Petroleum Gases

LEGISLATION

- Highly Flammable Liquids and Liquefied Petroleum Gases Regulations 1972 (SI 1972 No. 917) (HFL and LPG).

KEY DATES

- In force from 21/6/72 except Regulation concerning exhaust ventilation of workrooms which was in force from 21/6/74.

PURPOSE

- To protect employed persons from the fire and explosion risks associated with HFL and LPG.

APPROACH

- Lays down requirements for HFL and LPG including storage, labelling and fire precautions.

REGULATORY AUTHORITY

- HSE.

COVERAGE

- All factories where any HFL or LPG is present for the purposes of, or in connection with, any undertaking, trade or business.
- HFL is defined and includes liquids which support combustion and give off a flammable vapour at a temperature of less than 32 degrees Celsius.
- LPG means commercial butane, commercial propane and any mixture thereof.

COMPLIANCE REQUIREMENTS

- Requirements for HFL include:
  - Manner of storage and marking of both storage accommodation and vessels.
  - Precautions to be taken for the prevention of fire and explosion.
  - Provision, in certain cases, of fire fighting apparatus and a means of escape in case of fire.
- LPG must be stored, when not in use, according to a specified hierarchy with underground open air reservoirs the preferred option.
- LPG receptacles must be clearly and boldly marked with "Highly flammable – LPG" or similar. If this is impracticable this marking must be clearly and boldly displayed as near to the receptacle as possible.
- Employees have a duty to comply with such of the requirements of these Regulations which relate to doing, or not doing, acts by him, and use by him of plant, equipment or appliances.

DISCLOSURE

COMPLEMENTARY LEGISLATION

- Control of Major Accident Hazards Regulations 1999 (COMAH).
- Fire Certificates (Special Premises) Regulations 1996.

RELATED EC LEGISLATION

COMMENTS

RELATED DOCUMENTATION

- Testing results for HFL.

3.8 Lead

LEGISLATION

- Control of Lead at Work Regulations 2002 (SI 2002 No. 2676).

KEY DATES

- In force from 21/11/02.

PURPOSE

- Impose requirements for the protection of employees, and other persons, who might be exposed to lead at work.
- To reduce the exposure to lead of people at work.
- To implement EC Directives 82/605 and 98/24 (insofar as it relates to risks to health from exposure to lead).

APPROACH

- Require risk assessments to be conducted for any work with lead and take appropriate preventative or control measures.

REGULATORY AUTHORITY

- HSE.
- EHOs.

COVERAGE

- All employers (including self-employed) where any work activity may expose people to lead.
- Lead is defined and includes lead alkyls and alloys which are liable to be ingested or otherwise absorbed by persons. Lead from car exhausts is exempt from this definition.

COMPLIANCE REQUIREMENTS

- Only leadless or low solubility glazes (as defined) may be used in the manufacture of pottery.

- Employment of women of reproductive capacity or young persons is prohibited in certain processes *e.g.* lead–acid battery manufacture involving lead oxide manipulation.
- Employer shall not carry out work which is liable to expose employees to lead unless a suitable and sufficient assessment of the health risks has been undertaken including steps that need to be taken to meet these Regulations. These steps needed must be undertaken. Matters to consider in the risk assessment are specified and include level, type and duration of exposure, results of relevant medical surveillance and whether the exposure is significant. The risk assessment must be reviewed regularly and forthwith in certain circumstances including where there has been a significant change in the work. Where employer employs five or more employees the significant findings of the risk assessment and the steps to be taken shall be recorded.
- Exposure to lead should be prevented, or if this is not reasonably practicable, then adequately controlled where substitution shall be the preferred option. Protection measures shall be appropriate to the activity and in the following order of priority – design and use of appropriate processes, systems *etc.*, control of exposure at source and personal protective equipment (PPE). PPE is to be used as a last resort only. Respiratory protective equipment (RPE) may need to be provided.
- Employer shall take all reasonable measures to ensure any control measure is properly used or applied and employees shall make full and proper use of any control measure.
- Employers shall, so far as is reasonably practicable, ensure employees do not eat, drink or smoke in any place which is, or is liable to be, contaminated with lead.
- Control measures must be maintained in an efficient state, in efficient working order, in good repair and in a clean condition. Any local exhaust ventilation systems must be examined and tested at least every 14 months. Other engineering controls shall be examined and tested at suitable intervals.
- Concentrations of lead in air for significant exposures must be measured at least every three months.
- Employees exposed to lead, or liable to be so exposed, must be under a suitable medical surveillance programme where

exposure is significant, blood or urinary concentrations above certain levels or recommended by a doctor.

- Personal health records related to lead are required to be kept for at least 40 years from last entry.
- Where employees blood-lead concentrations exceed (or equal) certain specified action levels the employer must take action to reduce them so far as is reasonably practicable.
- Certain requirements are specified for biological monitoring *e.g.* for young persons and women of reproductive capacity this must be carried out at least every three months.
- Employees exposed to lead must be provided with appropriate and sufficient instruction and training.
- Where containers and pipes are not marked in accordance with other legislation then the employer shall ensure that the contents of those containers and pipes, together with the nature of those contents and any associated hazards, are clearly identifiable.
- Employer must prepare procedures, provide information and establish warning systems to deal with an emergency in the workplace related to the presence of lead.

DISCLOSURE

- The employer may not be entitled to the results of the biological tests because they are clinically confidential.

COMPLEMENTARY LEGISLATION

- Management of Health and Safety at Work (Amendment) Regulations 1994.
- Health and Safety (Young Persons) Regulations 1997.

RELATED EC LEGISLATION

- EC Directive 82/605 on the protection of workers from the risks related to exposure to metallic lead and its ionic compounds at work.

- EC Directive 98/24 on the protection of the health and safety of workers from risks related to chemical agents at work.

COMMENTS

- Reference should be made to COP2 *Control of Lead at Work*.
- Lead is a significant risk to young persons and women of reproductive capacity and therefore these Regulations must be considered with the legislation listed above under complementary legislation.
- These Regulations replaced the Control of Regulations at Work Regulations 1998 and introduced duties relating to emergency provisions. Any record keeping requirement under the 1998 Regulations shall remain in force as though these Regulations had not been passed.

RELATED DOCUMENTATION

- Biological monitoring results.
- Employees, exposed to lead, medical and training records.
- Air concentration results.
- Risk assessment.
- Maintenance records of control equipment *e.g.* PPE, RPE, local exhaust ventilation.

3.9 New Substances

LEGISLATION

- Notification of New Substances Regulations 1993 (SI 1993 No. 3050).
- Notification of New Substances (Amendment) Regulations 2002 (SI 2002 No. 2176).

KEY DATES

- In force from 31/1/94.
- 2002 Regulations in force from 17/9/02.

PURPOSE

- To ensure adequate information is known about new substances so that they can be used safely.
- Implement EC Directive 92/32 amending for the 7th time EC Directive 67/448 relating to the classification, packaging and labelling of dangerous substances.

APPROACH

- Before a new substance is supplied or imported a technical dossier must be sent to the HSE including, if it is a dangerous substance, appropriate classification, labelling and a safety data sheet.

REGULATORY AUTHORITY

- HSE.

COVERAGE

- New substances means any substance except a substance listed in the European Inventory of Existing Commercial Chemical Substances (EINECS).

- Substances that are placed on the market which is defined as supplying or making it available to another person within the Communities and includes importation of the substance.
- Intermediates means chemical substances which are solely manufactured for and consumed in or used for chemical processing in order to be transformed into another chemical substance.
- Certain substances are exempt usually because they are covered by other legislation, *e.g.* substances used as an active ingredient for a medicinal product as defined in the Medicines Act 1968, substances for use in food within the meaning of the Food Safety Act 1990, a radioactive substance within the meaning of the Ionising Radiations Regulations 1999.

COMPLIANCE REQUIREMENTS

- Must send to the HSE a full notification prior to placing a new substance on the market in a total quantity of one tonne or more per year. The new substance may not be placed on the market until 60 days after receipt of a conforming notification by the HSE. This notification includes:
  - A technical dossier supplying information necessary to evaluate foreseeable risk to human health and the environment including at least results of specified tests (identity, information on the substance (production, exposure, recipients, waste *etc.*), physico-chemical properties, toxicological studies, ecotoxicological studies, possibility of rendering substance harmless).
  - Certificate from body conducting tests stating conducted in accordance with good laboratory practice.
  - Declaration concerning unfavourable effects.
  - If substance is classified as dangerous – classification, labelling and safety data sheet.
  - An import statement if appropriate.
  - A statement requesting exemption from animal testing for one year. Only applicable if certain specified conditions are satisfied including that the substance has already been notified by another person. This is to avoid duplication of animal testing.

- Must inform HSE when quantity reaches specified amounts (first milestone from 10 tonnes per year or 50 tonnes in total manufacture). HSE may then require further testing to be undertaken.
- Must submit to HSE a reduced notification for substances placed on the market in quantities of less than one tonne per annum. Information and data required is specified but is less than for a full notification. The new substance may not be placed on the market until 30 days after receipt of a conforming notification by the HSE. For new substances placed on the market in quantities below 100 kg per year a notification is required but the data and information is even less onerous and is only required in summary form. The time period prior to placing on the market is 15 days.
- Substances placed on the market in quantities less than 10 kg per year per manufacturer do not require notification.
- Must comply with special provisions relating to polymers.
- Must inform HSE of any material change in new substance *e.g.* new knowledge of adverse effects, new uses.
- Must comply with certain import requirements for substances manufactured outside the European Community.
- Must pay the appropriate fees.
- Intermediates must be notified but a reduced test package is specified for intermediates, with limited exposure, above the 1 tonne threshold. Certain conditions must be satisfied including maximum number of two users' sites, substance must be rigorously contained by technical means during its whole lifecycle and labelled with "Caution – substance not yet fully tested".

DISCLOSURE

- Notification.

COMPLEMENTARY LEGISLATION

- Medicines Act 1968.
- Food Safety Act 1990.
- Ionising Radiations Regulations 1999.

- Notification of New Substances (Amendment) Regulations 2001 (NONS 2001).
- Biocidal Products Regulations 2001.
- Chemicals (Hazard Information and Packaging for Supply) Regulations 2002.

RELATED EC LEGISLATION

- EC Directive 92/32 amending for the 7th time EC Directive 67/448 relating to the classification, packaging and labelling of dangerous substances.
- EC Directive 2001/59 adapting to technical progress for the 28th time EC Directive 67/548 on the approximation of the laws, regulations and administrative provisions relating to the classification, packaging and labelling of dangerous substances.

COMMENTS

- A copy of the notification is sent to the European Commission.
- Provision is made for certain information sent in a notification to be confidential. A full justification must be given and the HSE will decide whether to keep that information confidential.
- NONS 2001 in force from 13th April 2001 exempted biocidal products, as defined in the Biocidal Products Regulations 2001, from the requirements of these Regulations.
- NONS 2002 provides for the notification of intermediates and specifies the tests required for a reduced test package.
- Reference could be made to *Making Sense of NONS (A Guide to the Notification of New Substances Regulations 1993)* HSE Books 1994.

RELATED DOCUMENTATION

- Test data including methods.
- Substance information including analytical methods.
- Production records.
- Technical dossier.
- Notification and associated correspondence.
- Information on use of substance.

3.10 Radioactive Substances

LEGISLATION

- Ionising Radiations Regulations 1999 (SI 1999 No. 3232).

KEY DATES

- In force from 1/1/00 except Regulation 5 (concerning carrying out of specified practises requiring HSE approval) in force from 13/5/00.

PURPOSE

- Protect all persons against ionising radiation arising from work with radioactive substances and other sources of ionising radiation.
- Lays down basic safety standards for public and employees against dangers of ionising radiation.

APPROACH

- The Regulations place a strict duty on employers to protect employees and other persons against ionising radiation arising from work and to notify the HSE of an intention to carry out work for the first time with such radiation. Specified practices require prior authorisation from the HSE before work can commence.
- The Regulations require risk assessment, specify arrangements for management of radiation protection, specify and control designated areas, require monitoring of certain persons, make arrangements for the control of radioactive substances and impose certain duties on employees.

REGULATORY AUTHORITY

- HSE.

COVERAGE

- Regulations apply to radiation employers, defined as employers who in the course of trade, business or other undertaking carry out or intend to carry out work with ionising radiation.
- Radioactive substance means any substance which contains one or more radionuclides whose activity cannot be disregarded for radiation protection.
- Ionising radiation means the transfer of energy in the form of particles or electromagnetic waves of a wavelength of 100 nanometres or less, or a frequency of 3×10^{15} hertz, or more, capable of producing ions directly or indirectly.
- Employer for the purposes of these Regulations includes the self-employed.

COMPLIANCE REQUIREMENTS

- A radiation employer shall not carry out specified practices, including the use of accelerators (except electron microscopes) and use of electrical equipment to produce X-rays for purpose of research, unless a prior written authorisation has been obtained from the HSE.
- A radiation employer must give at least 28 days notice to the HSE of the intention to carry out work with ionising radiation for the first time. Any material change to the nature of this work must be notified to the HSE.
- A radiation employer must carry out a risk assessment to assess hazards and risks to persons prior to any new work involving ionising radiation.
- Employer must comply with general duty to minimise exposure to ionising radiation as far as is reasonably practicable. Limits are specified on the doses of ionising radiation which persons may receive.
- Personal protective equipment (PPE) is to be used as a last resort only. Any engineering controls, PPE or safety controls must comply with certain standards and be adequately maintained and tested.
- A radiation employer must prepare a contingency plan, if a radiation accident is reasonably foreseeable, designed to restrict exposure if such an accident occurs.

- A radiation employer must consult a radiation protection adviser (RPA) for specified matters including examination and testing of engineering controls and other safety features.
- A radiation employer must ensure that adequate information, instruction and training is given to his employees and other persons who are engaged in work with ionising radiation.
- A radiation employer is required to follow specified requirements for designated areas which include areas where any person working is likely to receive an effective dose greater than 6 mSv a year.
- Employer must make arrangements to assess significant doses of ionising radiation received by employees, who are classified persons (*i.e.* likely to receive an effective dose in excess of 6 mSv per year or an equivalent dose which exceeds three-tenths of any relevant dose limit), with an approved dosimetry service (ADS), which is required to send summaries of all current dose records to HSE within three months of the end of each calendar year.
- A radiation employer must notify HSE and conduct an investigation if any person has received an overexposure as a result of work carried out by them.
- ADS must, when requested by employer and when a person ceases to be an employee, send to the HSE a termination record (which is a copy of the dose assessment relating to that person) for that employee.
- ADS must maintain and keep records relating to individual employees for at least 50 years.
- Employers shall ensure that certain specified persons including classified persons are subject to medical surveillance.
- A radiation employer must immediately notify the HSE where specified quantities of radioactive substances (in Schedule 8 of the Regulations) are released into the atmosphere or spilled so as to cause significant contamination.
- A radiation employer must ensure that radioactive substances are kept in suitable receptacles in a suitable store when they are not for the time being in use or being moved or disposed of.
- Certain requirements are specified for the control of radioactive substances, articles and equipment including sealed sources should be used where practicable.

- Records of quantities and locations of ionising radiation must be kept for 2 years from disposal.

DISCLOSURE

- HSE may request additional information.

COMPLEMENTARY LEGISLATION

- Radioactive Substances Act 1993.
- Ionising Radiation (Medical Exposure) Regulations 2000.
- Radiation (Emergency Preparedness and Public Information) Regulations 2001.

RELATED EC LEGISLATION

- EC Directive 90/641/Euratom on the operational protection of outside workers exposed to the risk of ionising radiation during their activities in controlled areas.
- EC Directive 96/29/Euratom laying down basic safety standards for the protection of the health of workers and the general public against the dangers arising from ionising radiation.
- EC Directive 97/43/Euratom on health protection of individuals against the dangers of ionising radiation in relation to medical exposure.

COMMENTS

- Non-ionising radiation, *e.g.* lasers are not covered by these Regulations but by general duties under the Health and Safety at Work *etc.* Act 1974 and the Management of Health and Safety at Work Regulations 1999.
- Ionising Radiation (Medical Exposure) Regulations 2000 mainly in force from 13th May 2000 lay down basic measures for the health protection of individuals (persons undergoing medical exposure whether as part of their own medical diagnosis or treatment or as part of occupational health surveillance, health screening, voluntary participation in research or medico-legal

procedures) against dangers of ionising radiation in relation to medical exposure.

- Certain exemptions are provided in the Regulations and explained in *Health and Safety Executive Statement on the Type Approval of Apparatus Under the Ionising Radiations Regulations 1999*.

RELATED DOCUMENTATION

- Notifications to HSE.
- Dose records/termination records.
- Records of quantities and locations of ionising radiation.
- Emergency contingency plan.
- Training records.

3.11 Safety Advisers

LEGISLATION

- Transport of Dangerous Goods (Safety Advisers) Regulations 1999 (SI 1999 No. 257).

KEY DATES

- In force from 1/3/99 except Regulation 4 (duty to appoint safety advisers) from 31/12/99.

PURPOSE

- Implement EC Directive 96/35 and ensure dangerous goods are transported safely.

APPROACH

- Prohibit the transport of dangerous goods by an employer unless he has appointed a suitably qualified safety adviser.
- Requires the safety adviser to perform certain duties.

REGULATORY AUTHORITY

- HSE.

COVERAGE

- Dangerous substances as defined in the Carriage of Dangerous Goods (Classification, Packaging and Labelling) and Use of Transportable Pressure Receptacles Regulations 1996 (see section on these Regulations).
- Apply to a self-employed person as they apply to an employer.
- Numerous exemptions are specified to cover certain circumstances, *e.g.* transport involving live animals, radioactive material in the body of any person, and an employer whose main or secondary activity is not the transport of dangerous goods and

who occasionally engages in the transport of dangerous goods and this transport does not create a significant risk to the health and safety of persons and to the environment.

COMPLIANCE REQUIREMENTS

- No employer shall transport dangerous goods by road, railway or inland waterway unless he has appointed a safety adviser for the purposes of advising him as to health, safety and environmental matters in connection with the transport of those goods.
- An employer must provide the safety adviser with adequate information and facilities to enable him to perform his function.
- Where more than one safety adviser is appointed employer must make arrangements for ensuring adequate cooperation between the safety advisers.
- The minimum functions of the safety adviser are specified in Schedule 2 to the Regulations and include monitoring compliance with the rules governing the transport of dangerous goods and preparing an annual report to the employer concerning the transport of dangerous goods.
- A safety adviser shall prepare a report of any accident which affects the health or safety of any person or causes damage to the environment or property.
- Keep the annual report and any accident reports for a minimum of 5 years.
- Provide the name of the safety adviser, accident or annual reports to the Secretary of State when requested to do so.

DISCLOSURE

- Name of safety adviser and certificate.
- Annual report.
- Accident report.

COMPLEMENTARY LEGISLATION

- Carriage of Dangerous Goods (Classification, Packaging and Labelling) and Use of Transportable Pressure Receptacles Regulations 1996.

RELATED EC LEGISLATION

- EC Directive 96/35 on the appointment and vocational qualification of safety advisers for the transport of dangerous goods by road, rail and inland waterway.

COMMENTS

- The safety adviser must hold a valid certificate issued by or on behalf of the Secretary of State. A certificate will only be issued where the safety adviser has completed training and passed an examination. The training and examination must include certain specified matters.
- The HSE can grant exemption to any person or classes of person provided this does not compromise the health and safety of persons or the environment.

RELATED DOCUMENTATION

- Safety adviser training records and certificate.
- Annual report and associated documentation.
- Accident report and associated documentation.
- Any relevant audit of facilities.
- Procedures for handling and transporting dangerous goods.

4 PLANT AND EQUIPMENT

4.1 Batteries and Accumulators

LEGISLATION

- Batteries and Accumulators (Containing Dangerous Substances) Regulations 1994 (SI 1994 No. 232).
- Batteries and Accumulators (Containing Dangerous Substances) Amendment Regulations 2000 (SI 2000 No. 3097) and 2001 (SI 2001 No. 2551).

KEY DATES

- 1994 regulations in force from 1/3/94 except Regulation 4 (marking requirements) from 1/8/94.
- 2000 Regulations in force from 18/12/00.
- 2001 Regulations in force from 31/8/01.

PURPOSE

- Ensure batteries containing heavy metals are used safely.
- Implement EC Directives 91/157, 93/86 and 98/101.

APPROACH

- Bans the marketing of certain batteries and accumulators.
- Require certain markings to be on batteries or accumulators.

REGULATORY AUTHORITY

- Secretary of State.

COVERAGE

- Batteries and accumulators providing a source of electrical energy generated by direct conversion of chemical energy and consisting of one or more primary (non-rechargeable) batteries or secondary (rechargeable) cells.

- Batteries and accumulators put on the market as from 1st January 1999 containing more than 0.0005% of mercury by weight.
- Batteries and accumulators put on the market as from 18th September 1992 and containing:
  - More than 25 mg of mercury per cell (except alkaline manganese batteries).
  - More than 0.025% of cadmium by weight.
  - More than 0.4% of lead by weight.
- Alkaline manganese batteries containing more than 0.025% of mercury by weight placed on the market as from 18th September 1992.

COMPLIANCE REQUIREMENTS

- Must not market a battery or accumulator that is prohibited. These are ones that contain more than 0.0005% of mercury by weight, other than button cells, or button cells with a mercury content of no more than 2% by weight.
- Must mark the battery or accumulator with specified information including collection mark and the relevant heavy metal content mark.
- Must ensure battery or accumulator can be easily removed when spent if they are incorporated into appliances (obligation on manufacturer of appliance).

DISCLOSURE

COMPLEMENTARY LEGISLATION

RELATED EC LEGISLATION

- EC Directive 91/157 on batteries and accumulators containing more than specified levels of dangerous substances (mercury, cadmium or lead) as amended by EC Directive 98/101.

- EC Directive 93/86 providing symbols to indicate the separate collection of such batteries and accumulators as well as the relevant chemical symbol to indicate their heavy metal content.

COMMENTS

- Secretary of State may give notice that Regulations are not being complied with and specify matters to be undertaken to achieve compliance.
- Provision is made for individual criminal liability where offence is committed by a body corporate with the consent or connivance of, or to be attributable to any neglect on the part of specified individuals within the body corporate.
- 2000 Regulations amended the definition of batteries and accumulators to which the Regulations apply.
- 2001 Regulations amended a defect in the 2000 Regulations. The effect is to remove an exception for batteries and accumulators produced in or imported into the European Economic Area before the Directive came into force.

RELATED DOCUMENTATION

- Battery and accumulator composition details.
- Battery and accumulator purchase requests and any manufacturer information on them.
- Any Secretary of State notice.

4.2 Cooling Towers and Evaporative Condensers

LEGISLATION

- Notification of Cooling Towers and Evaporative Condenser Regulations 1992 (SI 1992 No. 2225).

KEY DATES

- In force from 2/10/92.

PURPOSE

- Ensure that local authorities are aware of cooling towers and evaporative condensers within their geographical area.

APPROACH

- Places various duties on persons who have to any extent control of premises to notify the local authority:
  - Where a cooling tower or an evaporative condenser is situated on those premises.
  - Where any changes to the information required under the original notification occurs.
  - Where the tower or condenser ceases to be, and is no longer intended to remain, a notifiable device (as defined).

REGULATORY AUTHORITY

- Local authority.

COVERAGE

- Regulations apply to all non-domestic premises used for, or in connection with, the carrying on of a trade, business, or other undertaking (for profit or not) on which a notifiable device is situated.

- Notifiable device means a cooling tower or an evaporative condenser, except where:
  - It contains no water that is exposed to air.
  - Its water supply is not connected; and
  - Its electrical supply is not connected.
- "Cooling tower" is defined as a device whose main purpose is to cool water by direct contact between that water and a stream of air.
- "Evaporative condenser" is defined as a device whose main purpose is to cool a fluid by passing that fluid through a heat exchanger which is itself cooled by contact with water passing through a stream of air.

COMPLIANCE REQUIREMENTS

- Must notify the local authority in writing – on form approved by HSE – where notifiable device is situated on premises. That notification must contain the following information:
  - The address of the premises where the notifiable device is to be situated.
  - The name, address and telephone number of a person who has, to any extent, control of such premises.
  - The number of notifiable devices at the premises.
  - The location on the premises of each notifiable device.
- Where the premises are to any extent under the control of the manufacturer of the device, must notify the local authority simply of the address of the premises, and the name, address and telephone number of a person who has, to any extent, control of those premises.
- Must notify any changes to the information required under the above notifications in writing to the local authority concerned within one month of their occurrence.
- Must notify local authority as soon as is reasonably practicable where a device ceases to be, and is no longer intended to remain, a notifiable device (*N.B.* This requirement does not apply where the operation of a notifiable device is suspended for the purpose of maintenance or by reason of seasonal shutdown).

DISCLOSURE

COMPLEMENTARY LEGISLATION

- Health and Safety at Work *etc.* Act 1974.

RELATED EC LEGISLATION

COMMENTS

RELATED DOCUMENTATION

- Notifications.
- Any notifications of changes.
- Any notification of cessation.

4.3 Display Screen Equipment

LEGISLATION

- Health and Safety (Display Screen Equipment) Regulations 1992 (SI 1992 No. 2792) (DSE Regulations).

KEY DATES

- In force from 1/1/93.

PURPOSE

- To ensure use of DSE occurs with minimum risk to health and safety of users *e.g.* avoiding musculo-skeletal problems, visual fatigue and mental stress.
- To enact into UK law EC Directive 90/270.

APPROACH

- Requires employers to identify any hazards and then to evaluate risks to health and safety for work with DSE.
- Establish minimum health and safety requirements for work with DSE by adopting ergonomic principles.

REGULATORY AUTHORITY

- HSE.
- EHOs.

COVERAGE

- Employed persons at work who are users and self-employed persons who are occupiers.
- Exemptions include DSE in driver's cabs or control cabs for vehicles or machinery, on board a means of transport, mainly intended for public operation, portable systems not in prolonged use, calculators, cash registers and window typewriters.

- User is defined as an employee who habitually uses DSE as a significant part of his normal work. An operator is the same but for self-employed people (see comments).

COMPLIANCE REQUIREMENTS

- Employers must ensure that any workstation which may be used for the purposes of his undertaking meets specified requirements with regards to equipment, *e.g.* tilting chair, software, *e.g.* suitability, and environment, *e.g.* lighting glare. Assessments must be regularly reviewed.
- Additional requirements relate to the daily routine of users, information and training.
- Employers are obliged to provide, at their expense, eye and eyesight tests in response to a request from an employee who is a user.

DISCLOSURE

COMPLEMENTARY LEGISLATION

- Health and Safety (Miscellaneous Amendments) Regulations 2002.

RELATED EC LEGISLATION

- EC Directive 90/270 on the minimum safety and health requirements for work with DSE.

COMMENTS

- Originally the requirements concerning equipment only applied to users and operators but this restriction was removed by the Miscellaneous Amendments 2002 Regulations. This was done to give full effect to EC Directive 90/270.
- The definition of user in the Regulations is given above but official guidance expands on this. It states a user is a worker for

whom use is more or less continuous on most days. Typical examples are categorised into definite, possible or non-users. For a user most or all of the following criteria apply:
  - Job cannot be done effectively or at all without DSE.
  - Worker has no discretion over whether to use DSE.
  - Job requires significant training or particular skills.
  - The worker uses DSE for periods of an hour or more at a time, more or less on a daily basis.
  - The task depends upon transfer of information between the worker and screen.
  - Attention and concentration demands are high, such as where there may be critical consequences of an error.
- The Regulations are general in nature and reference should be made to L26 *DSE Work – Guidance on Regulations* for more details.
- An assessment need not be recorded if, for example, no significant risks are identified or the workstation is temporary. Information provided by the user (after having received adequate training) is an essential part of the assessment.

RELATED DOCUMENTATION

- DSE assessment.
- Information from users (*e.g.* in the form of a checklist).
- Information and training records.

4.4 Electricity

LEGISLATION

- Electricity at Work Regulations 1989 (SI 1989 No. 635).

KEY DATES

- In force from 1/4/90.

PURPOSE

- To ensure the safety of electrical equipment at work and safety during work on electrical systems.

APPROACH

- Specify general standards for the design, construction, maintenance and operation of electrical equipment and systems at work.

REGULATORY AUTHORITY

- HSE.
- EHOs.

COVERAGE

- The use of all electrical equipment (as defined in the Regulations) at work.

COMPLIANCE REQUIREMENTS

- All electrical systems must, so far as is reasonably practicable, be constructed and maintained so as to prevent danger at all times.
- Any protective equipment must be suitable for the intended use, well maintained and properly used.

- Standards are specified for a wide range of activities including earthing precautions, integrity of referenced conductors, connections, work on or near live conductors, work on dead equipment.
- Must ensure the competence and necessary technical knowledge of staff who undertake any electrical work.
- Routine portable appliance testing and labelling will sometimes be necessary.

DISCLOSURE

COMPLEMENTARY LEGISLATION

RELATED EC LEGISLATION

COMMENTS

- The onus of the Regulations is on the employer to assess work activities which utilise electricity and to define foreseeable risks and minimise them.
- The Regulations impose the application of general health and safety principles to specific electrical activities.
- HSE may grant written exemptions if satisfied this will not prejudice health and safety of persons affected.

RELATED DOCUMENTATION

- Service records for electrical equipment.
- Any electricity diagram for buildings *etc*.

4.5 Explosive Atmospheres

LEGISLATION

- Equipment and Protective Systems Intended for Use in Potentially Explosive Atmospheres Regulations 1996 (SI 1996 No. 192).
- Equipment and Protective Systems Intended for Use in Potentially Explosive Atmospheres (Amendment) Regulations 2001 (SI 2001 No. 3766).

KEY DATES

- 1996 Regulations in force from 1/3/96.
- 2001 Amendment Regulations in force from 21/12/01.

PURPOSE

- Implement, in the UK, EC Directive 94/9 on the approximation of the laws of Member States concerning equipment and protective systems intended for use in potentially explosive atmospheres (ATEX Directive).
- To ensure equipment and protective systems used in potentially explosive atmospheres are safe.

APPROACH

- Imposes health and safety standards for appropriate equipment and protective systems.

REGULATORY AUTHORITY

- HSE.

COVERAGE

- Equipment and protective systems intended for use in potentially explosive atmospheres, devices and components. Each of these phrases is widely defined in detail in Regulation 3.

- Certain specified equipment is exempt including domestic and non-commercial equipment, where potentially explosive atmospheres may only be rarely created as a result of an accidental leak, and medical devices.

COMPLIANCE REQUIREMENTS

- A responsible person (manufacturer, manufacturer's authorised representative or the person who places product on the market), who places on the market equipment, a protective system or device, has a duty to ensure that it satisfies the relevant essential health and safety requirements and that the appropriate conformity assessment procedure has been carried out. The conformity assessment procedure is set out in Schedules 6 to 12 and the appropriate procedure is determined on the basis of equipment group and category of the product.
- Must affix CE (European Approved Standard) mark to appropriate product and the product must be safe.
- Responsible person who places a component on the market must ensure the appropriate conformity assessment procedure has been carried out and that it is accompanied by a certificate.
- Must pay fees.

DISCLOSURE

- Conformity assessment test results.

COMPLEMENTARY LEGISLATION

- Electrical Equipment for Explosive Atmospheres (Certification) Regulations 1990 as amended.

RELATED EC LEGISLATION

- EC Directive 94/9 on the approximation of the laws of Member States concerning equipment and protective systems intended for use in potentially explosive atmospheres.

COMMENTS

- Transitional provisions exist whereby the Regulations do not apply to equipment, or a protective system, placed on the market on or before 30th June 2003 which complies with the health and safety provisions in respect of it which were in force in Great Britain on 23rd March 1994.
- Electrical Equipment for Explosive Atmospheres (Certification) Regulations 1990 provide the standards for certification of appropriate explosive atmosphere equipment which apply to equipment covered by the transitional provisions. These Regulations and all the amendments will be revoked from 1st July 2003 (at the end of the transitional period).
- Regulations provide for the appointment of notified bodies with specified duties including ensuring manufacturer fulfils obligations arising out of the relevant quality assurance procedure.
- Provision is made for liability of persons other than the principal offender.
- 2001 Amendment Regulations make typographical changes and include requirements relating to the activity of "putting into service" relevant equipment in specific situations.

RELATED DOCUMENTATION

- Conformity assessment test results and procedures.
- CE mark and any associated documentation.
- Manufacturing procedures and specifications for equipment.

4.6 Lifting – Mechanical

LEGISLATION

- Lifting Operations and Lifting Equipment Regulations 1998 (SI 1998 No. 2307) (LOLER).

KEY DATES

- In force from 5/12/98.

PURPOSE

- To ensure safety during the use of lifting equipment.
- To implement the lifting requirements of EC Directive 95/63 into UK law.

APPROACH

- Establish required standards for lifting equipment.

REGULATORY AUTHORITY

- HSE.
- EHOs.

COVERAGE

- LOLER applies to all premises where the Health and Safety at Work *etc*. Act 1974 applies, *e.g*. factories, schools and universities, agricultural premises and homeworkers.
- Applies to self-employed.
- Lifting equipment means work equipment for lifting or lowering loads and includes its attachments used for anchoring, fixing or supporting it. This is a very wide definition and includes a wide range of equipment, *e.g*. cranes, hoists and lifts.

COMPLIANCE REQUIREMENTS

- Employers shall ensure lifting equipment is of adequate strength and stability for the load.
- Employers have certain duties relating to lifting equipment used for lifting persons *e.g.* to ensure such equipment prevents a person using it being crushed, trapped or struck or falling from the carrier.
- Employers shall ensure that lifting equipment is positioned or installed so as to minimise, as far as is reasonably practicable, risks of the load striking a person and the load drifting or falling unintentionally.
- Employers shall ensure machinery and accessories for lifting loads are clearly marked to indicate safe working loads (or safe working loads for each configuration of the machine).
- Employers shall ensure every lifting operation involving lifting equipment is properly planned by a competent person, appropriately supervised and carried out in a safe manner.
- Employers must comply with specified duties for inspection and examination of lifting equipment. This includes examinations for defects prior to first use and continued examinations during use with specified minimum re-examination periods for certain lifting equipment.
- A person making a thorough examination of lifting equipment shall forthwith notify the employer of any defects if it is or could become dangerous. A report must be compiled and sent to the employer. If the defect is such as to make the lifting equipment liable to cause serious personal injury a copy of the report shall also be sent to the HSE. In both cases the lifting equipment shall not be used until the defect is fixed.

DISCLOSURE

- Defect reports to HSE.

COMPLEMENTARY LEGISLATION

- Health and Safety at Work *etc.* Act 1974.
- Personal Protective Equipment at Work Regulations 1992.

- Provision and Use of Work Equipment Regulations 1998 (PUWER).
- Management of Health and Safety at Work Regulations 1999.
- Health and Safety (Miscellaneous Amendments) Regulations 2002.

RELATED EC LEGISLATION

- EC Directive 95/63 amending EC Directive 89/655 on the minimum health and safety requirements for the use of equipment by workers at work.

COMMENTS

- LOLER requirements need to be considered in conjunction with other legislation. PUWER provides general requirements, *e.g.* suitability, for all work equipment. The Management of Health and Safety at Work Regulations 1999 provides that risk assessments must be carried out for all work activities including lifting operations and provide requirements related to what activities persons under 18 can undertake.
- For more details reference should be made to L113 *Safe Use of Lifting Equipment: Lifting Operations and Lifting Equipment Regulations 1998: ACOP and Guidance.*
- Miscellaneous Amendments regulations in force from 17th September 2002 made minor drafting changes to LOLER including correcting errors.

RELATED DOCUMENTATION

- Defect reports.
- Examination reports including thorough examinations.
- Training records of operators.
- Equipment examiners records.
- Equipment maintenance records.

4.7 Personal Protective Equipment – Head Protection

LEGISLATION

- Construction (Head Protection) Regulations 1989 (SI 1989 No. 2209).

KEY DATES

- In force from 30/3/90.

PURPOSE

- To prevent head injuries on construction sites.

APPROACH

- Require the provision, maintenance and use of suitable head protection during construction work.

REGULATORY AUTHORITY

- HSE.

COVERAGE

- Apply to building operations and works of engineering construction within, in either case, the meaning of the Factories Act 1961. In practice this means most construction activities.

COMPLIANCE REQUIREMENTS

- Suitable head protection should be designed to provide protection, as far as is reasonably practicable, against foreseeable risks to head injuries.
- Employers must provide each employee with suitable head protection while at work (on a site to which the Regulations apply) and shall maintain it, and replace it when necessary.

- Employers shall ensure, so far as is reasonably practicable, that head protection is worn unless there is no foreseeable risk of injury to the head other than by falling.
- Self-employed persons must provide, maintain and replace their own head protection.
- Any written rules regulating the wearing of head protection must be brought to the notice of relevant people.
- Employees shall wear head protection when required to do so by employer and notify employer of any loss of, or obvious defect in the head protection.

DISCLOSURE

COMPLEMENTARY LEGISLATION

- Factories Act 1961.
- Employment Act 1989.

RELATED EC LEGISLATION

COMMENTS

- Exemption from these Regulations is provided for Sikhs wearing turbans by the Employment Act 1989.
- The HSE can provide a written exemption if satisfied that the health and safety of persons who are likely to be affected will not be prejudiced because of it.

RELATED DOCUMENTATION

- Written rules or procedures.

4.8 Personal Protective Equipment – Work

LEGISLATION

- Personal Protective Equipment at Work Regulations 1992 (SI 1992 No. 2966) (PPE Regulations).

KEY DATES

- In force from 1/1/93.

PURPOSE

- To ensure that the need for PPE is identified and acted upon.
- To ensure PPE is suitable and in working order.

APPROACH

- Require an assessment of risks that have not been adequately controlled by other means to identify the need for PPE. PPE is to be used as a last resort only.
- Employers must provide suitable PPE to protect employees and maintain such PPE.

REGULATORY AUTHORITY

- HSE.
- EHOs.

COVERAGE

- All work activities except where more detailed legislation on specific matters covers PPE (see complementary legislation which lists the exceptions).
- PPE as defined includes equipment, *e.g.* eye protectors, and clothing, *e.g.* aprons, which are intended to be worn or held and which protect against risks to health and safety.

COMPLIANCE REQUIREMENTS

- Prior to supplying PPE the employer must assess and ensure that the proposed PPE is suitable.
- Employers must provide suitable PPE to protect employees from risks that are not adequately controlled by other means. Self-employed must make provision for themselves for their own activities.
- PPE must be appropriate for risk involved, conditions where it is to be worn and the period for which it is worn.
- PPE must take account of ergonomic principles and the state of health of person wearing it.
- PPE must be provided for use by only one person where it is necessary to ensure PPE is hygienic and otherwise free of risk to health.
- PPE must be maintained in an efficient state and appropriate storage must be provided for PPE when not in use.
- Employers must have adequate arrangements to give information, instruction and training on PPE including use and risks protecting against.
- Employers shall, where appropriate and at suitable intervals, organise demonstrations in the wearing of PPE.
- Employers should take all reasonable steps to ensure PPE is properly used.
- Employee shall forthwith report any loss of, or defect to, PPE.

DISCLOSURE

COMPLEMENTARY LEGISLATION

- Road Traffic Act 1988.
- Construction (Head Protection) Regulations 1989.
- Noise at Work Regulations 1989.
- Health and Safety (Display Screen Equipment) Regulations 1992.
- Ionising Radiations Regulations 1999.
- Control of Asbestos at Work Regulations 2002.
- Control of Lead at Work Regulations 2002.

- Control of Substances Hazardous to Health Regulations 2002.
- Personal Protective Equipment Regulations 2002.
- Health and Safety (Miscellaneous Amendments) Regulations 2002.

RELATED EC LEGISLATION

- EC Directive 89/656 on the minimum health and safety requirements for the use by workers of PPE at the workplace (PPE Directive).
- EC Directive 89/686 on the approximation of the laws of the Member States relating to personal protective equipment as amended by EC Directives 93/68, 93/95 and 96/58.

COMMENTS

- Reference should be made to L25 *Personal Protective Equipment at Work Regulations 1992: Guidance on Regulations*.
- The Personal Protective Equipment Regulations 2002 in force from 15th May 2002 implement EC Directive 89/686 and prescribe standards for manufacturers and suppliers of PPE. These include that basic specified health and safety requirements applicable to the particular PPE have been satisfied. Also the PPE must have a CE mark affixed to it and the procedure and requirements to obtain this mark are detailed.
- Miscellaneous Amendments Regulations 2002 in force from 17th September 2002 introduced the requirement for demonstrations and various other changes which are reflected in the Compliance section.

RELATED DOCUMENTATION

- PPE records.
- Employee records of PPE use and issue.

4.9 Pressure Systems

LEGISLATION

- Pressure Systems Safety Regulations 2000 (SI 2000 No. 128).

KEY DATES

- In force from 21/2/00.

PURPOSE

- To ensure the safety of pressure systems.
- To prevent certain vessels from becoming pressurised.

APPROACH

- Impose specific safety requirements for pressure systems which are used or intended to be used at work.
- Impose safety requirements to prevent certain vessels from becoming pressurised.

REGULATORY AUTHORITY

- HSE.

COVERAGE

- Pressure systems which are used or intended to be used at work.
- Pressure systems are defined and include a system comprising one or more pressure vessels of rigid construction, any associated pipework and protective devices.
- Numerous exceptions are listed including weapon systems, certain parts of rail mounted vehicles, vehicle tyres and portable fire extinguishers.
- Covers self-employed person in respect of their own activities at work.

COMPLIANCE REQUIREMENTS

- Any person who designs, manufactures, imports or supplies any pressure system shall ensure that the following are complied with:
  - It is properly designed and constructed from suitable material so as to prevent danger.
  - It is designed and constructed so that all necessary examinations for preventing danger can be carried out.
  - If access is available to its interior, it shall be designed and constructed so that access can be gained without danger.
  - It is provided with such protective devices as may be necessary for preventing danger.
- Designer or supplier of pressure systems must provide sufficient written information to enable the Regulations to be complied with.
- Installer must ensure pressure system is installed so that nothing about the way it is installed gives rise to danger.
- Manufacturer of pressure vessel must ensure certain stated information is marked on the vessel or a plate attached to it. This includes manufacturer's name, serial number to identify the vessel, maximum allowable pressure of the vessel and date of manufacture of the vessel.
- Must not operate a pressure system unless the safe operating limits have been established.
- User of a pressure system must have a written scheme for the periodic examination, by a competent person, of specified parts of the system. Must comply with requirements of the written examination report.
- If written examination report identifies imminent danger unless certain repairs, modifications or changes are carried out then the pressure system must not be operated until such work is satisfactorily completed. The competent person shall send the report to the enforcing authority within 14 days of completing the examination.
- Must provide suitable instruction for safe operation of system and action to be taken in emergency to operators of the pressure system.
- Must comply with specified record keeping requirements.

DISCLOSURE

- Imminent danger report.

COMPLEMENTARY LEGISLATION

RELATED EC LEGISLATION

COMMENTS

- HSE may provide exemptions from Regulations, subject to conditions, and only if satisfied that the health and safety of persons who are likely to be affected by the exemption will not be prejudiced as a consequence of it.

RELATED DOCUMENTATION

- Imminent danger report.
- Installation records.
- Maintenance records.
- Examination records.
- Insurance examination reports if relevant.
- Operators training records.
- Designers or suppliers written documents.

4.10 Transportable Pressure Vessels

LEGISLATION

- Transportable Pressure Vessels Regulations 2001 (SI 2001 No. 1426).

KEY DATES

- In force from 1/7/01 except Regulations concerning interpretation, citation, commencement and fees in force from 3/5/01.

PURPOSE

- To ensure transportable pressure vessels are placed on the market only after taking safety into account.
- To ensure transportable pressure vessels are used safely at work.
- Implement EC Directive 96/36 concerning the approximation of laws of member States relating to common provision for transportable pressure equipment and methods for inspection.

APPROACH

- Transportable pressure vessels can be placed on the market or used at work only after certain safety standards have been complied with and assessed by a competent body.

REGULATORY AUTHORITY

- HSE.

COVERAGE

- Transportable pressure vessels defined as a cylinder, tube or cryogenic receptacle (including certain valves or accessories) which is used or intended to be used for the storage or transport of a gas or certain other specified liquids, *e.g.* stabilised hydrogen cyanide.

- Numerous exemptions are listed in Schedule 1 and include gas cylinders used as a component part of breathing appliances, certain fire extinguishers and transporting gases contained in foodstuffs or beverages.
- Placing on the market means supplying or making it available to another person in the Communities and includes importation.
- Applies to any transportable pressure vessel used at work and manufactured on or after 1st July 2001 and on or before 30th June 2001 where that vessel is subject to a reassessment of conformity or an EEC type cylinder.

COMPLIANCE REQUIREMENTS

- No person shall place on the market or use at work a transportable vessel (including valves and other accessories with a safety function) unless it is:
  - Safe and suitable for purpose.
  - Designed, manufactured and tested according to specified standards.
  - Assessed by a notified or approved body and bears any suitable markings (requirements differ between vessels used exclusively within Great Britain and those which are not).
- Permits the reassessment for conformity of transportable pressure vessels manufactured prior to 1st July 2003 which do not bear the conformity marking.
- Must ensure that the transportable pressure vessel is periodically inspected in accordance with specified procedures to ensure safety of that vessel and that it continues to meet the standards and requirements to which it was designed and manufactured.
- Any required marking must be visible, easily legible and indelible.
- Must pay fees.

DISCLOSURE

- Notified body disclosure.
- Approved body disclosure.

COMPLEMENTARY LEGISLATION

- Carriage of Dangerous Goods (Classification, Packaging and Labelling) and Use of Transportable Pressure Receptacles Regulations 1996 (CDGCPL).
- Pressure Equipment Regulations 1999.

RELATED EC LEGISLATION

- EC Directive 76/767 concerning the approximation of laws of the Member States relating to common provision for pressure vessels and methods for inspecting them.
- EC Directives 84/525, 84/526 and 84/527 concerning the approximation of laws of the Member States relating to seamless steel gas cylinders, seamless unalloyed aluminium and aluminium alloy gas cylinders, and welded unalloyed steel gas cylinders, respectively.
- EC Directive 99/36 concerning the approximation of laws of the member States relating to common provision for transportable pressure equipment and methods for inspection.

COMMENTS

- Until 1st July 2003, for transportable pressure vessels manufactured on or after 1st July 2001, any person may choose to comply with CDGCPL Regulations or these Regulations.
- General safety requirements and reassessment of conformity provisions do not apply to an EEC-type cylinder.
- Certain defences are specified including that the offence was committed by the act or default of another person not being an employee of the offending party and that the offending party took all reasonable precautions and exercised all due diligence to avoid commission of the offence.

RELATED DOCUMENTATION

- Transportable pressure vessels documentation – service records, manufacturing details, design *etc.*

- Conformity testing results and correspondence.
- Notification records.
- Inspection records.

4.11 Work Equipment

LEGISLATION

- Provision and Use of Work Equipment Regulations 1998 (SI 1998 No. 2306) (PUWER).

KEY DATES

- In force from 5/12/98 except certain requirements for existing mobile work equipment which came into force on 5/12/02.

PURPOSE

- To ensure that work equipment should not result in health and safety risks, regardless of its age or origin.
- To implement EC Directives 89/655 and 95/63.

APPROACH

- Establish minimum requirements for the selection and use of work equipment.

REGULATORY AUTHORITY

- HSE.
- EHOs.

COVERAGE

- Work equipment means any machinery, appliance, apparatus, tool or installation for use at work and includes mobile and lifting equipment. This is a very wide definition and covers almost any equipment used at work *e.g.* hammers, photocopiers and laboratory apparatus.
- PUWER applies to all premises where the Health and Safety at Work *etc.* Act 1974 applies, *e.g.* factories, schools and universities, and to homeworkers and premises where domestic staff are employed, *e.g.* kitchens of sheltered accommodation.
- Applies to self-employed.

COMPLIANCE REQUIREMENTS

- Employers must ensure that all work equipment is constructed, or adapted, to be suitable for its purpose and is only used for operations for which it is suitable.
- Employers must ensure work equipment is maintained in an efficient state, in efficient working order and in good repair. Any maintenance log must be kept up to date.
- Employers must comply with requirements relating to inspection, instructions and training, *e.g.* must ensure adequate health and safety information and written instructions, where appropriate, are available to all work equipment users.
- Employers shall ensure that work equipment conforms, at all times, with any essential requirements. "Essential requirements" means requirements relating to the design and construction of work equipment including product legislation, *e.g.* Supply of Machinery (Safety) Regulations 1992 and carries a CE (European Approved Standard) mark if appropriate.
- Employers must provide adequate safety arrangements with regard to the physical aspects of work equipment including guards, stop controls, stability, lighting, markings and warnings.
- Employers must prevent exposure to a health and safety risk to persons from specified hazards, *e.g.* work equipment catching fire, or adequately control the risk where prevention is not reasonably practicable. Personal protective equipment should be used as a last resort only.
- Employers have certain duties related to the following physical attributes of work equipment:
  - Guards.
  - High or very low temperatures.
  - Controls for starting or making a significant change in operation.
  - Controls and control systems including emergency and non-emergency stop controls.
  - Stability.
  - Lighting and maintenance.
  - Markings and warnings.

- Employers must comply with certain duties for mobile work equipment *e.g.* minimising any risk to an employee from the equipment rolling over.
- Employer must comply with specified management duties for power presses.

DISCLOSURE

COMPLEMENTARY LEGISLATION

- Health and Safety at Work *etc.* Act 1974.
- Health and Safety (Display Screen Equipment) Regulations 1992.
- Personal Protective Equipment at Work Regulations 1992.
- Supply of Machinery (Safety) Regulations 1992.
- Workplace (Health, Safety and Welfare) Regulations 1992.
- Reporting of Injuries, Diseases and Dangerous Occurrences Regulations 1995 (RIDDOR).
- Construction (Health, Safety and Welfare) Regulations 1996.
- Confined Spaces Regulations 1997.
- Lifting Operations and Lifting Equipment Regulations 1998 (LOLER).
- Management of Health and Safety at Work Regulations 1999.
- Control of Substances Hazardous to Health Regulations 2002 (COSHH).
- Health and Safety (Miscellaneous Amendments) Regulations 2002.

RELATED EC LEGISLATION

- EC Directive 89/655 concerning the minimum safety and health requirements for the use of work equipment by workers at work.
- EC Directive 95/63 amending Directive 89/655 on the minimum health and safety requirements for the use of equipment by workers at work.

COMMENTS

- PUWER revokes specific legislation concerning abrasive wheels, power presses and woodworking machines. Certain specific requirements for power presses are included in PUWER but the principle is that you evaluate and manage the risks these machines present in the same way as any work equipment.
- EC Directive 95/63 contains lifting requirements which are enacted into UK law by LOLER but PUWER still applies. LOLER imposes additional requirements specific to the risks associated with lifting operations and equipment.
- PUWER is closely linked to other legislation specified above in complementary legislation, *e.g.* the Management of Health and Safety at Work Regulations 1999 imposes a requirement to conduct risk assessments and this will include assessments relating to the use of work equipment.
- PUWER is general in nature and reference should be made to L22 *Safe Use of Work Equipment: Provision and Use of Work Equipment Regulations 1998: ACOP and Guidance.*
- Miscellaneous Amendments Regulations 2002 in force from 17th September 2002 made amendments to give clearer effect to conformity with Community requirements by introducing and defining essential requirements.

RELATED DOCUMENTATION

- Any maintenance log.
- Risk assessments.
- Machine operating instructions and associated manufacturers machine documentation.
- Employee training records.
- Any written operating instructions.

5 PREMISES

5.1 Construction

LEGISLATION

- Construction (Design and Management) Regulations 1994 (SI 1994 No. 3140) (CONDAM).
- Construction (Design and Management) (Amendment) Regulations 2000 (SI 2000 No. 2380).

KEY DATES

- CONDAM in force from 31/3/95.
- Amendment Regulations in force from 2/1/01.

PURPOSE

- To improve safety during construction work.

APPROACH

- Define the parties involved and allocate safety management responsibilities to them.
- Requires the preparation of a health and safety file and plan.

REGULATORY AUTHORITY

- HSE.

COVERAGE

- All construction work including maintenance, renovation and repair where the work will last for more than 30 days or will involve more than 500 person days.

- Note: Regulations always apply to (a) design and demolition work regardless of the duration of the activity and (b) non-notifiable construction work where five or more people are on site at any one time.
- Occupied offices, shops, restaurants *etc.* where minor construction work is being carried out are exempt.

COMPLIANCE REQUIREMENTS

- Client must select and appoint a competent planning supervisor and principal contractor (who can be the same person).
- Planning supervisor must:
  - Ensure that the safety aspects of the project design are addressed so far as is reasonably practicable.
  - Ensure cooperation of different designers in relation to health and safety.
  - Prepare a health and safety file for each structure in the project, maintain this file and deliver to the client at the end of the project.
  - Prepare a pre-tender health and safety plan containing specified information.
  - Ensure the HSE is notified of the project.
- Client must ensure health and safety file is made available for inspection by any person who may need it to comply with statutory obligations.
- The designer should ensure that the health and safety of those who are going to construct, maintain or repair a structure should be considered during the design process.
- The principal contractor must:
  - Ensure coordination and cooperation of all contractors on site.
  - Ensure, so far as is reasonably practicable, all contractors and employees comply with the health and safety plan.
  - Ensure any required notifications to HSE concerning the project are displayed on site.
- Contractors have a duty to cooperate with and provide relevant health and safety information to the principal contractor.

- Contractors and other parties must contribute appropriate information to the safety file.
- Safety plan must be made available to the principal contractor.
- Safety file must be handed over to the client.

DISCLOSURE

- Notification to HSE.

COMPLEMENTARY LEGISLATION

- Construction (Health, Safety and Welfare) Regulations 1996.

RELATED EC LEGISLATION

- EC Directive 92/57 on the implementation of minimum safety and health requirements at temporary or mobile construction sites.

COMMENTS

- The Regulations were to give effect to the UK's obligations to enact EC Directive 92/57 but differences still exist *e.g.* the Directive specifies a wider range of particulars to be included in the safety plan than CONDAM does.
- The Amendment Regulations make minor changes to the definition of designer and include people working for the designer under his control.

RELATED DOCUMENTATION

- Tender documents.
- Construction contract.
- Safety file.
- Safety plan.
- Any risk assessments concerning the construction project.
- Any notification to HSE.

5.2 Construction Premises

LEGISLATION

- Construction (Health, Safety and Welfare) Regulations 1996 (SI 1996 No. 1592).

KEY DATES

- In force from 2/9/96.

PURPOSE

- To improve the health, safety and welfare of persons at work on construction sites.
- To implement certain requirements of EC Directive 92/57.

APPROACH

- Imposes requirements to reduce the risk from each of the main hazards in the construction industry.

REGULATORY AUTHORITY

- HSE.
- EHOs.
- Fire authority (for requirements concerning fire matters).

COVERAGE

- Construction work is extensively defined and means the carrying out of any building, civil engineering or engineering contruction work. The Regulations specify activities which are included *e.g.* maintenance, demolition, repair of a structure.
- Employers, self-employed and anyone carrying out or in control of construction work must comply with the Regulations.

COMPLIANCE REQUIREMENTS

- Employers carrying out construction work must take suitable and sufficient steps (sometimes as far as is reasonably practicable) to:
  - Provide a safe place of work.
  - Prevent falls, including persons falling, persons falling through fragile objects and objects falling.
  - Ensure structures are stable and excavations do not collapse.
  - Ensure no person is at risk from the use of explosives.
  - Maintain cofferdams and caissons and ensure the design and construction materials are appropriate for the use.
  - Prevention or avoidance of drowning.
  - Ensure traffic matters are organised without risks to health.
  - Prevent fire including provision of emergency routes and exits, emergency procedures, fire detection and fire fighting equipment.
- Employers must ensure suitable and sufficient arrangements for dealing with foreseeable emergencies on construction sites including designated persons to implement the arrangements and necessary contact with external services.
- Employers must ensure that certain welfare standards are met including fresh air, temperature, weather protection, lighting, and cleanliness requirements.
- Employers shall ensure any person who carries out any activity involving construction work has appropriate training, knowledge or experience to enable them to carry out that work with minimum risk of injury to persons.
- Employers must comply with inspection and report requirements for specified places of work.

DISCLOSURE

COMPLEMENTARY LEGISLATION

- Construction (Design and Management) Regulations 1994 (CONDAM).

- Lifting Operations and Lifting Equipment Regulations 1998 (LOLER).
- Management of Health and Safety at Work Regulations 1999 (amends Regulations).

RELATED EC LEGISLATION

- EC Directive 92/57 on the implementation of minimum safety and health requirements at temporary or mobile construction worksites.

COMMENTS

- LOLER applies lifting requirements which must be complied with when lifting equipment is used including use on construction sites.

RELATED DOCUMENTATION

- Any CONDAM safety files or plans.
- Site inspection reports.
- Tender contract.

5.3 Safety Signs and Signals

LEGISLATION

- Health and Safety (Safety Signs and Signals) Regulations 1996 (SI 1996 No. 341).

KEY DATES

- In force from 1/4/96 except for fire safety signs lawfully in use before that date in force from 24/12/98.

PURPOSE

- To ensure safety signs and signals are clear and easily understood.
- Implement EC Directive 92/58 – the Safety Signs Directive.

APPROACH

- Specifies minimum requirements and standards for safety signs and signals.

REGULATORY AUTHORITY

- HSE.
- EHOs.

COVERAGE

- All premises and activities except signs specified for supply and carriage of dangerous goods, and signs used to control road, rail, air, sea and inland waterway traffic.

COMPLIANCE REQUIREMENTS

- Employers must provide and maintain safety and/or warning signs where the risk assessment required under the Management of Health and Safety at Work Regulations 1999 identifies that the risk cannot be controlled by other means.

- Employers must provide suitable and sufficient instructions and training regarding safety signs.
- Required safety signs must comply with relevant requirements and standards specified in the Regulations including specified colours and symbols.
- Certain specific requirements must be observed for specified articles, *e.g.* containers, pipes, fire fighting equipment and illuminated signs.
- Hand signals must be precise, easy to carry out and understand – acceptable hand signals are given in the Regulations.

DISCLOSURE

COMPLEMENTARY LEGISLATION

- Management of Health and Safety at Work Regulations 1999.

RELATED EC LEGISLATION

- EC Directive 92/58 on the minimum requirements for the provision of safety and/or health signs at work (Safety Signs Directive).

COMMENTS

- The main types of sign are specified in the Regulations as follows:
  - Prohibition, *e.g.* no smoking.
  - Warning, *e.g.* danger electricity.
  - Mandatory, *e.g.* ear protection must be worn.
  - Emergency escape or first aid signs.
  - Fire fighting.
- Reference should be made to L64 *Safety Signs and Signals: Guidance on Regulations* for details.

RELATED DOCUMENTATION

- Risk Assessments.

5.4 Workplace

LEGISLATION

- Workplace (Health, Safety and Welfare) Regulations 1992 (SI 1992 No. 3004) (Workplace Regulations).

KEY DATES

- In force from 1/1/93 (for new workplaces, modifications, extensions or conversions).
- In force from 1/1/96 (for existing workplaces).

PURPOSE

- To ensure that workplaces meet the health, safety and welfare needs of each member of the workforce including disabled people.
- Implement EC Directive 89/654.

APPROACH

- Impose various minimum requirements in relation to workplaces including maintenance of the workplace.
- Requirements are prescribed by standards, *e.g.* lighting, ventilation *etc.* or assessments, *e.g.* traffic flow.

REGULATORY AUTHORITY

- HSE.
- EHOs/local authorities.

COVERAGE

- Regulations requirements apply to all persons in a "workplace", meaning any non-domestic premises, or part of premises, which are made available to any person as a place of work.

- Exceptions include construction sites, ships and sites where mineral exploration or extraction is undertaken, and the Regulations' application is modified in relation to temporary work sites, specified means of transport and specified agricultural or forestry undertakings.
- Disabled person has the meaning given by Section 1 of the Disability Discrimination Act 1995.

COMPLIANCE REQUIREMENTS

- Employers must ensure that adequate health, safety and welfare standards are achieved in relation to the following:
  - Working environment including ventilation, heating and lighting.
  - Work space, including cleanliness, room dimensions and work stations.
  - Safety, including condition of floors, design and operation of windows, storage arrangements and the cleaning of windows.
  - Traffic arrangements, including pedestrians, vehicles, construction of doors, gates and escalators.
  - Welfare facilities, including toilets, washing facilities, cloakrooms, drinking water and facilities for rest and to eat meals.
- Where a workplace is in a building, the building shall have a stability and solidity appropriate to the nature of the use of the workplace.
- Rest rooms and areas shall include suitable arrangements to protect non-smokers from the discomfort caused by tabacco smoke and an adequate number of tables and seats.
- Parts of workplace, used or occupied by disabled persons at work, shall be organised to take account of such persons.

DISCLOSURE

COMPLEMENTARY LEGISLATION

- Docks Regulations 1988.
- Electricity at Work Regulations 1989.

- Personal Protective Equipment at Work Regulations 1992.
- Health and Safety (Display Screen Equipment) Regulations 1992.
- Disability Discrimination Act 1995.
- Construction (Health, Safety and Welfare) Regulations 1996 (which amends the Workplace Regulations).
- Provision and Use of Work Equipment Regulations 1998 (PUWER).
- Management of Health and Safety at Work Regulations 1999.
- Control of Substances Hazardous to Health Regulations 2002 (COSHH).
- Health and Safety (Miscellaneous Amendments) Regulations 2002.

RELATED EC LEGISLATION

- EC Directive 89/654 concerning the minimum safety and health requirements for the workplace (Workplace Directive).

COMMENTS

- The Regulations apply to a very wide range of workplaces, not only factories, shops and offices but also to schools, hospitals and places of entertainment.
- Reference should be made to the Health and Safety Commission's Approved Code of Practice and Guidance L8 *Workplace Health, Safety and Welfare*.
- These Regulations did not enact the fire safety provisions of the Workplace Directive. The fire provisions are implemented by the Fire Precautions (Workplace) Regulations 1997.
- Forklift truck operations are governed by a separate ACOP.
- Miscellaneous Amendments Regulations 2002 in force from 17th September 2002 introduced requirements including those for disabled persons and rest rooms.

RELATED DOCUMENTATION

- Site/building plans.
- Maintenance records.
- Equipment records.
- Assessments.

6 EMPLOYEES

6.1 Confined Spaces

LEGISLATION

- Confined Spaces Regulations 1997 (SI 1997 No. 1713).

KEY DATES

- In force from 28/1/98.

PURPOSE

- To ensure work is undertaken safely in confined spaces.

APPROACH

- Requires elimination of working in confined spaces or if unavoidable a safe system of work must be operated.

REGULATORY AUTHORITY

- HSE.
- EHOs.

COVERAGE

- Confined space means any place in which, by virtue of its enclosed nature, there arises a reasonably foreseeable risk of serious injury arising from fire or explosion, loss of consciousness, drowning or asphyxiation.
- The Regulations apply to all premises and work subject to the Health and Safety at Work *etc.* Act 1974 except diving operations, below ground in a mine and certain activities on a ship.
- Employers, any person to any extent under employers' control and self-employed have to comply with the Regulations.

COMPLIANCE REQUIREMENTS

- No person shall enter a confined space to work for a purpose unless it is not reasonably practicable to achieve that purpose without such entry.
- No person shall work in a confined space unless a system of work is in place, which renders the work safe, and without risks to health.
- No person shall enter a confined space unless adequate arrangements have been made to undergo, if necessary, a safe rescue.

DISCLOSURE

COMPLEMENTARY LEGISLATION

- Health and Safety at Work *etc.* Act 1974.
- Personal Protective Equipment at Work Regulations 1992.
- Construction (Design and Management) Regulations 1994 (CONDAM).
- Provision and Use of Work Equipment Regulations 1998 (PUWER).
- Management of Health and Safety at Work Regulations 1999.
- Control of Asbestos at Work Regulations 2002.
- Control of Lead at Work Regulations 2002.
- Control of Substances Hazardous to Health Regulations 2002 (COSHH).

RELATED EC LEGISLATION

COMMENTS

- Certain specific requirements related to other hazards, *e.g.* lead and hazardous substances are covered by the legislation specified above under complementary legislation but apply equally to confined spaces where the relevant hazard is present.
- A permit-to-work system is not compulsory by the Regulations but it may be that such a system demonstrates a safe system of work if it is operated efficiently and complied with.

- Further details are available in L101 *Safe Work in Confined Spaces: Confined Spaces Regulations 1997: ACOP, Regulations and Guidance.*

RELATED DOCUMENTATION

- Relevant risk assessments.
- Procedure for permit-to-work in confined spaces (and permits) if operated.
- Rescue procedure and arrangements.

6.2 Employee Consultation – With Employees

LEGISLATION

- Health and Safety (Consultation with Employees) Regulations 1996 (SI 1996 No. 1513).

KEY DATES

- In force from 1/10/96.

PURPOSE

- To clarify consultation rights of employees who are not members of a recognised trade union.
- Implement certain requirements of EC Directive 89/391 for unrepresented employees.

APPROACH

- Require employers to consult with unrepresented employees on health and safety matters.

REGULATORY AUTHORITY

- HSE.
- EHOs.

COVERAGE

- Employees who are not represented by a safety representative.

COMPLIANCE REQUIREMENTS

- Employers must consult with employees, not represented by a safety representative, on matters relating to their health and safety at work, including the consequences of introducing any new technology.

- Employers may consult with either the employees directly or through employee representatives elected by a group of employees to represent them in consultations on health and safety matters.
- Information provided to employees or representatives must enable them to participate fully and effectively in consultations. Specified disclosure exemptions are stated, *e.g.* if it would be against the national interest.
- Employee representative functions are specified including making representations to employer on potential hazards which could affect the group of employees he represents.
- Provision is provided for training, time off and facilities for employee representatives and candidates.

DISCLOSURE

COMPLEMENTARY LEGISLATION

- Safety Representatives and Safety Committees Regulations 1977.

RELATED EC LEGISLATION

- EC Directive 89/391 on the introduction of measures to encourage improvements in the safety and health of workers at work (Framework Directive).

COMMENTS

- The requirement for these Regulations resulted from a judgement of the European Court which decided that the UK had not fully implemented the Framework Directive with regards to these workers.
- Reference should be made to L87 *Safety Representatives and Safety Committees (Rev) – A Guide to Health and Safety (Consultation with Employees) Regulations 1996.*

DOCUMENTATION

- Safety information provided.
- Training records of employee representatives.

6.3 Employee Consultation – Information for Employees

LEGISLATION

- Health and Safety Information For Employees Regulations 1989 (SI 1989 No. 682).
- Health and Safety Information For Employees (Modifications and Repeals) Regulations 1995 (SI 1995 No. 2923).

KEY DATES

- 1989 Regulations in force from 18/10/89.
- 1995 Regulations all in force by 1/1/98.

PURPOSE

- To provide all employees with specified health and safety information.

APPROACH

- Requires employer to display a poster, or give a leaflet, with specified information to all employees.

REGULATORY AUTHORITY

- HSE.
- Environmental Health Departments of local authorities.

COVERAGE

- All employers and employees except the master and crew of a sea-going ship.

COMPLIANCE REQUIREMENTS

- Employers shall ensure that either a poster approved by the HSE is in a place that is reasonably accessible to all employees or give all employees an approved leaflet. The poster or leaflet must include the name of the enforcing authority for the premises and the address of the employment medical advisory service for the area.

DISCLOSURE

- Displaying the approved poster.
- Providing employees with approved leaflet.

COMPLEMENTARY LEGISLATION

RELATED EC LEGISLATION

COMMENTS

- The 1995 Regulations provided that the HSE can approve posters or leaflets which are specific to particular classes of employment so that employers may display or provide those specific posters or leaflets rather than ones of a general nature.

RELATED DOCUMENTATION

6.4 First Aid

LEGISLATION

- Health and Safety (First Aid) Regulations 1981 (SI 1981 No. 917).

KEY DATES

- In force from 1/7/82.

PURPOSE

- To enable employees to receive first aid from qualified first aiders while at work if they are injured or get ill.

APPROACH

- Require employers to provide adequate provision with regards to first aid.

REGULATORY AUTHORITY

- HSE.
- Environmental Health Departments of local authorities.

COVERAGE

- First aid facilities for employees and the self-employed who are injured or become ill at work.
- All premises except certain mines, offshore installations and domestic workplaces.

COMPLIANCE REQUIREMENTS

- Employers must provide adequate and appropriate equipment and facilities to enable first aid to be rendered to his/her employees if required. The nature of the undertaking and number of employees are two important factors in determining whether any provision is adequate and appropriate.

- Employers must provide the number of first aiders as is adequate and appropriate, and the first aiders must have undergone approved HSE training and re-training.
- Employers must inform their employees of the arrangements made in connection with the provision of first aid.
- Self-employed person shall ensure such equipment is available to enable him to render first aid to himself.
- Any first-aid room provided shall be easily accessible to stretchers and similar equipment and be sign-posted.

DISCLOSURE

COMPLEMENTARY LEGISLATION

- Health and Safety (Miscellaneous Amendments) Regulations 2002.

RELATED EC LEGISLATION

COMMENTS

- 2002 Miscellaneous Amendment Regulations in force from 17th September 2002 added the requirements concerning first-aid rooms.
- The Regulations are general in nature and reference should be made to L74 *First Aid at Work Approved Code of Practice and Guidance Notes* for more details.
- Note: The Regulations and ACOP do not provide a formula for number of first aiders required.

RELATED DOCUMENTATION

- Training records of first aiders.

6.5 Genetically Modified Organisms

LEGISLATION

- Genetically Modified Organisms (Contained Use) Regulations 2000 (SI 2000 No. 2831).

KEY DATES

- In force from 15/10/00.

PURPOSE

- To protect persons and the environment from risks arising from activities involving the contained use of genetically modified microorganisms (GMMO).
- To protect persons from risks arising from activities involving the contained use of genetically modified organisms (GMO) which are not microorganisms.

APPROACH

- Prohibit any activity involving GMMOs and GMOs unless a risk assessment has been undertaken.
- Impose requirements relating to notifying a competent authority.
- Establish certain safety and containment standards.

REGULATORY AUTHORITY

- HSE.
- For England and Wales – the Secretary of State and the Minister for the Environment, Food and Rural Affairs.
- For Scotland the Scottish Ministers.

COVERAGE

- Genetic modification is defined as the altering of the genetic material in an organism in a way that does not occur naturally.
- Contained use – definition includes culturing, storing, transporting *etc*.
- Includes educational establishments and for the purposes of these Regulations a student shall be treated as an employee.

COMPLIANCE REQUIREMENTS

- Must conduct a risk assessment for risks to human health and the environment before any activity involving genetic modification of microorganisms is undertaken. The matters to take into account are specified and include any potential harmful effects, their severity and likelihood. The steps to be followed are stated and include identification of harmful effects and the characteristics of the environment likely to be exposed.
- Must conduct a risk assessment for risks to human health before any activity involving genetic modification of organisms other than microorganisms is undertaken. The matters to take into account are specified and include any potential harmful effects, their severity and likelihood. The steps to be followed are stated and include identification of harmful effects of the recipient and the donor organism and adjustment of the level of risk in the light of the matters found in the assessment.
- Risk assessments must be recorded and reviewed if no longer valid or their has been a significant change. Record must be kept for at least 10 years.
- Must establish a genetic modification safety committee if a risk assessment is undertaken.
- Must notify competent authority of proposed use of premises for an activity involving genetic modification prior to first use.
- Must give prior notification, together with certain information, to the competent authority of specified activities involving genetic modification and in stated circumstances the competent authority must give its consent.

- Must, if genetic modification undertaken, ensure certain specified safety principles are observed and appropriate containment measures (detailed in the Regulations) are selected.
- Must prepare, in certain circumstances, an emergency plan to secure the health of persons or the health of persons and the protection of the environment.
- Must report every accident and provide the competent authority with information about the accident.
- Must pay fees.

DISCLOSURE

- Notifications to competent authority.
- Accident reports and information.
- Register of notifications.

COMPLEMENTARY LEGISLATION

- Environmental Information Regulations 1992.
- Genetically Modified Organisms (Risk Assessment) (Records and Exemptions) Regulations 1996.
- Genetically Modified Organisms (Contained Use) Amendment Regulations 2002.
- Genetically Modified Organisms (Deliberate Release) Regulations 2002.

RELATED EC LEGISLATION

- EC Directive 90/219 on the contained use of genetically modified microorganisms as amended by EC Directives 94/51 and 98/81.
- EC Directive 2001/18 on the deliberate release into the environment of genetically modified organisms.

COMMENTS

- Provision is made for disclosure of information and publicity. This includes that the competent authority maintains a register of notifications. Copies of this register are kept in certain offices of the HSE. Certain information may be classified as confidential if certain criteria are satisfied and the onus is on the applicant to justify the confidentiality. The grounds for confidentiality are one or more as specified in the Environmental Information Regulations 1992. The 2002 Amendment Regulations in force from 8th February 2002 makes provision for keeping information confidential if it is in the interest of national security to do so.

- The competent authority may grant exemption to the Regulations but only if satisfied that the health and safety of persons and the environment are not prejudiced by the granting of such exemption.

RELATED DOCUMENTATION

- Notifications to competent authority and correspondence.
- Risk assessments.
- Safety committee minutes.
- Accident reports.
- Procedures for genetic manipulation.
- Containment and safety measures documentation.
- Emergency plan.

6.6 Manual Handling

LEGISLATION

- Manual Handling Operations Regulations 1992 (SI 1992 No. 2793).

KEY DATES

- In force from 1/1/93.

PURPOSE

- Apply modern ergonomic approach to prevent injury caused by manual handling.
- Implement EC Directive 90/269.

APPROACH

- Require employers to eliminate so far as is reasonably practicable, manual handling operations where there is a risk of injury.
- Require risk assessments to be undertaken.

REGULATORY AUTHORITY

- HSE.
- EHOs.

COVERAGE

- Manual handling is defined as all operations involving the moving or supporting of a load by hand or bodily force.
- Employees and self-employed – the latter have a duty to apply Regulations in respect of their own activities.
- Crew of a sea-going vessel under the direction of master is exempt.

COMPLIANCE REQUIREMENTS

- Factors are specified to determine whether manual handling at work involves a risk of injury and to determine the appropriate steps to take to reduce that risk. These are:
  - Physical suitability of the employee.
  - Clothing, footwear or other personal effects employee is wearing.
  - Employee knowledge and training.
  - Results of any relevant risk assessment.
  - Whether employee is member of a group particularly at risk.
  - Results of any health surveillance.
- A hierarchy of measures is established. Employers:
  - Must avoid hazardous manual handling operations, so far as is reasonably practicable.
  - Assess any hazardous manual handling operation that is unavoidable.
  - Reduce the risk of injury so far as is reasonably practicable.
- Assessments must be carried out by a competent person and regularly reviewed, *e.g.* if new information is available or a reportable injury occurs.
- Employers must provide employees with health and safety information and training, *e.g.* how to recognise a potentially hazardous handling operation, proper use of handling aids.
- Employees while at work are placed under certain duties, *e.g.* to use any system of work properly, to highlight any difficulties.

DISCLOSURE

COMPLEMENTARY LEGISLATION

- Reporting of Injuries, Diseases and Dangerous Occurrences Regulations 1995 (RIDDOR).
- Management of Health and Safety at Work Regulations 1999.
- Health and Safety (Miscellaneous Amendments) Regulations 2002.

RELATED EC LEGISLATION

- EC Directive 90/269 on the minimum health and safety requirements for the manual handling of loads where there is a risk in particular of back injury to workers.

COMMENTS

- 2002 Miscellaneous Amendments Regulations in force from 17th September 2002 introduced the factors to determine whether there is a risk of injury and what steps to take to reduce the risk.
- For more detailed guidance, reference should be made to L23 *Manual Handling: Guidance on Regulations.*
- Assessments should be recorded unless they are low risk or simple. The guidance gives extensive information on the types of assessments and a broad structure for detailed assessments. This structure takes into account tasks, loads, working environment, individual capability and other factors. Some of these factors have now been made law by the 2002 Amendments Regulations.

RELATED DOCUMENTATION

- Manual handling assessment.
- Information and training records.
- Relevant reportable accident and employee records.

6.7 Noise

LEGISLATION

- Noise at Work Regulations 1989 (SI 1989 No. 1790).

KEY DATES

- In force from 1/1/90.

PURPOSE

- Introduce provisions protecting workers from exposure to excessive noise at work.
- Implement EC Directive 86/188 on the protection of workers from the risks related to exposure to noise at work.

APPROACH

- Require employers to assess the exposure of their employees to noise and take necessary steps to reduce that exposure.
- Impose requirements on employers in relation to the marking of, and entry of employees into, ear protection zones.
- Impose requirements on employers to provide information, instruction and training to those employees likely to be exposed to specified noise levels.

REGULATORY AUTHORITY

- HSE.
- EHOs.
- Secretary of State for Defence.

COVERAGE

- All work activities where harmful noise is likely to be present.
- References to employers includes the self-employed for the purposes of these Regulations.

COMPLIANCE REQUIREMENTS

- Employers must ensure that noise assessment is carried out by competent person when employee's daily exposure to noise is likely to be or above 85 dB(A) or a level of peak sound pressure of 200 Pascals or above, and keep records of such assessments.
- Employers must reduce employee's exposure to noise so far as is reasonably practicable (other than by providing personal ear protectors) whenever their daily exposure to noise is likely to be or exceed 90 dB(A) or a level of peak sound pressure of 200 Pascals or above.
- Employers must provide suitable and efficient personal ear protectors to employees (at the employee's request) whenever their daily exposure to noise is likely to be or exceed 85 dB(A) but be less than 90 dB(A).
- Employers must provide suitable personal ear protectors to employees whenever their daily exposure to noise is likely to be or exceed 90 dB(A) or a level of peak sound pressure of 200 Pascals or above. Employees are under a duty to make full and proper use of the protectors in such circumstances.
- Employers must comply with requirements in relation to the marking of, and entry of employees into, ear protection zones (*i.e.* areas where employees daily exposure to noise is likely to be or exceed 90 dB(A) or a level of peak sound pressure of 200 Pascals or above).
- Employers must comply with requirements to maintain, in an efficient state and working order and in good repair, any personal ear protectors required to be provided and that such are properly used.
- Employers must comply with requirements to provide adequate information, instruction and training to those employees likely to be exposed to specified noise levels.

DISCLOSURE

COMPLEMENTARY LEGISLATION

- Personal Protective Equipment at Work Regulations 1992 (PPE).

- Provision and Use of Work Equipment Regulations 1998 (PUWER).
- Management of Health and Safety at Work Regulations 1999.

RELATED EC LEGISLATION

- EC Directive 86/188 on the protection of workers from the risks related to exposure to noise at work.

COMMENTS

- The Regulations make no reference to hearing tests. The need for hearing tests may be identified by risk assessments required under the Management of Health and Safety at Work Regulations 1999. Regulation 5 of these Regulations states that employers should provide such health surveillance as identified by the risk assessment.
- Any ear protectors provided shall comply with the standards specified in the PPE Regulations in so far as they relate to ear protectors.
- Provision is made for issue of exemption certificates by the HSE in relation to Regulation's requirements where this is not likely to prejudice the health and safety of the persons affected.
- Reference could be made to *The Noise at Work Regulations: a Brief Guide to the Requirements for Controlling Noise at Work (INDR(L)(Rev))* (reprinted November 1999 with textual amendments).

RELATED DOCUMENTATION

- Noise assessment records.
- Noise measurement records.
- Affected employees records.
- Relevant information provided to employees.

6.8 Pregnant Women

LEGISLATION

- Management of Health and Safety at Work Regulations 1999 (SI 1999 No. 3242) Regulations 16, 17 and 18.

KEY DATES

- In force from 29/12/99.

PURPOSE

- To encourage improvements in the safety and health at work of pregnant women and new mothers.

APPROACH

- Require particular attention to pregnant women when carrying out risk assessments and to place restrictions on certain types of work.

REGULATORY AUTHORITY

- HSE.
- Environmental Health Departments of local authorities.

COVERAGE

- Women of childbearing age for risk assessment requirement. Women known to be pregnant or new mothers for other provisions.

COMPLIANCE REQUIREMENTS

- The risk assessment required by Regulation 3 The Management of Health and Safety at Work Regulations 1999 shall, if persons working in the undertaking include women of childbearing age and if work involves physical, biological or chemical agents, include an assessment of risks to the pregnant mother and foetus.

- If the risk cannot be avoided, then the employees conditions shall be altered so as not to expose the employee to the risk. If it is unreasonable to move the employee to other work then the employee shall be suspended to avoid any risks.
- Where a new or expectant mother works at night and a doctor certifies that she should not do this then the employee shall be suspended.
- The employees conditions of employment shall not be altered unless the employee has notified the employer in writing that she is pregnant, has given birth within the previous six months or is breast-feeding.

DISCLOSURE

- Employee discloses pregnancy, breast-feeding or recently given birth to employer.

COMPLEMENTARY LEGISLATION

- Management of Health and Safety at Work Regulations 1999

RELATED EC LEGISLATION

- EC Directive 92/85 on the introduction of measures to encourage improvements in the safety and health at work of pregnant workers and workers who have recently given birth or are breast-feeding (*The Pregnant Workers Directive*).

COMMENTS

RELATED DOCUMENTATION

- Employees health records
- Risk assessments

6.9 Whistle-blowers

LEGISLATION

- Public Interest Disclosure Act 1998.
- Public Interest Disclosure (Compensation) Regulations 1999 (SI 1999 No. 1548).
- Public Interest Disclosure (Prescribed Persons) Order 1999 (SI 1999 No. 1549).

KEY DATES

- All in force from 2/7/99.

PURPOSE

- To encourage workers to report specified matters, including breaches of legislation and factors endangering the environment and/or health and safety of persons, to specified bodies.

APPROACH

- To protect the whistle-blower from adverse action by the employer if a protected disclosure is made.
- Amend (by adding new provisions) the Employment Rights Act 1996.
- Disclosures may be made to "prescribed persons" who are defined by the 1999 Order.

REGULATORY AUTHORITY

- HSE.
- Local authorities.
- Environment Agency.

COVERAGE

- Workers as widely defined are covered and include all employees.

COMPLIANCE REQUIREMENTS

- When workers make a protected disclosure it is unlawful for them to be subjected to any adverse action by their employer and any dismissal will be automatically unfair. The compensation in such a case is not subject to any statutory cap and is therefore unlimited.
- For a disclosure to be protected it needs to satisfy requirements concerning:
  - Subject matter (*e.g.* a criminal offence has been committed or the health and safety of any individual has been, or is likely to be, endangered, or the environment has been, is being, or is likely to be damaged).
  - Whom the information is disclosed to (specified in 1999 Order and includes the HSE and local authorities who are prescribed persons only for health and safety matters disclosed to them, and the Environment Agency for environmental matters).
  - Circumstances of the disclosure, *e.g.* disclosure is in good faith and the worker reasonably believes that the information or any allegation are substantially true.

DISCLOSURE

- Worker disclosure to HSE and/or local authority.
- Worker disclosure to the Environment Agency.

COMPLEMENTARY LEGISLATION

- Employment Rights Act 1996.

RELATED EC LEGISLATION

COMMENTS

- The whistle-blower is not protected if disclosure is itself illegal.
- The workers remedy, if adverse action follows a prescribed disclosure, is complaint to an employment tribunal.

- It is advisable for employers to make provision and have procedures for reporting of environmental and health and safety matters by ALL employees.

RELATED DOCUMENTATION

- Environment and safety committee minutes and documentation.
- Any environmental or health and safety document concerning the prescribed disclosure.

6.10 Working Time

LEGISLATION

- Working Time Regulations 1998 (SI 1998 No. 1833) as amended by the Working Time Regulations 1999 (SI 1999 No. 3372).

KEY DATES

- In force from 1/10/98.
- 1999 regulations in force from 17/12/99.

PURPOSE

- To protect employees from exploitation by employers and thereby minimise work related stress and tiredness.
- Implement EC Directive 93/104 concerning certain aspects of the organisation of working time.

APPROACH

- Provides minimum terms and conditions of employment in relation to working time.

REGULATORY AUTHORITY

- HSE.
- Local authorities.

COVERAGE

- Every employee except those working in specified transport sectors, sea fishing, domestic servants and doctors in training.
- Exemptions are provided from certain requirements for some groups *e.g.* armed forces and family workers.
- Exemption is provided from requirements, concerning weekly working time, night work, rest periods and breaks, for workers who work the duration of which is not measured, predetermined or can be determined by the worker himself. If only part

of the workers work time satisfies this condition then weekly working time and night work requirements only apply to the workers working time that is measured, predetermined or cannot be determined by the worker himself.

COMPLIANCE REQUIREMENTS

- A worker's working time shall not exceed an average of 48 hours in seven days unless the worker has agreed in writing to waive this right. An employer shall keep up-to-date records for such workers.
- An employer shall take all reasonable steps to ensure that a night worker's normal hours of work do not exceed an average of eight hours for each 24 hours.
- An employer shall not assign an adult worker to night work unless the employee has had the opportunity of a free health assessment. Similar provision exists for workers between the ages of 15 and 18. The health assessment shall be repeated at appropriate regular intervals.
- If a doctor advises the employer of health problems suffered due to night work by a particular worker then the employer shall transfer the worker to non-night work if it is possible.
- If the pattern of work puts a workers health and safety at risk, in particular because work is monotonous or the work-rate is pre-determined then the employer shall ensure the worker is given adequate rest breaks.

DISCLOSURE

COMPLEMENTARY LEGISLATION

- Employment Rights Act 1996.

RELATED EC LEGISLATION

- EC Directive 93/104 concerning certain aspects of the organisation of working time (*Working Time Directive*).
- EC Directive 94/33 on the protection of young people at work (*Young Workers Directive*).

COMMENTS

- These Regulations are made under The Employment Rights Act 1996 and contain provisions which have a more contractual than health and safety element *e.g.* entitlements to paid annual leave. These requirements are individual worker rights and enforced by proceedings before employment tribunals.
- The Regulations provide that the HSE is responsible for enforcing certain requirements (those listed under compliance requirements above) and make breach of these requirements a criminal offence.
- Originally the HSE had express powers specified in the Regulations to inspect workers records but this power was removed by the 1999 amendment Regulations.

RELATED DOCUMENTATION

- Work time records.
- Health assessment records.

6.11 Young Persons

LEGISLATION

- Management of Health and Safety at Work Regulations 1999 (SI 1999 No. 3242) Regulation 19.

KEY DATES

- In force from 29/12/99.

PURPOSE

- To provide additional health and safety protection for young persons at work.
- To implement certain requirements from EC Directive 94/33.

APPROACH

- Provide additional health and safety requirements for employers who employ young persons.

REGULATORY AUTHORITY

- HSE.
- EHOs.

COVERAGE

- Young person is a person who has not reached the age of 18.
- Child means a person who is not over compulsory school age as defined in the Education Act 1996.

COMPLIANCE REQUIREMENTS

- Employers must carry out a risk assessment (or amend an existing one) before taking young persons into their employment. Specified factors must be taken into account *e.g.*

inexperience, immaturity, lack of awareness of risks to their health and safety, any exposure to biological, chemical or physical agents.

- Employers must provide the parent or guardian of a child with sufficient health and safety information before the child is employed including any risks identified and associated preventative/control measures.
- Employers may not employ young persons in specified circumstances such as where the work is beyond their physical or psychological capacity, or involves harmful exposure to radiation.

DISCLOSURE

- Information provided to parent or guardian.

COMPLEMENTARY LEGISLATION

- Education Act 1996.
- Management of Health and Safety at Work Regulations 1999.

RELATED EC LEGISLATION

- EC Directive 94/33 on the protection of young people at work (Young workers Directive).

COMMENTS

- Reference should be made to HS(G)165 *Young People at Work* for more details.

RELATED DOCUMENTATION

- Birth certificates.
- Risk assessments.
- Information and correspondence with parent or guardian of a child.

B ENVIRONMENT

7 INTEGRATED POLLUTION CONTROL

7.1 Integrated Pollution Prevention and Control (IPPC) and Local Air Pollution Prevention and Control (LAPPC)

LEGISLATION

- Pollution Prevention and Control Act 1999.
- Pollution Prevention and Control (England and Wales) Regulations 2000 (SI 2000 No. 1973) (PPCR).
- Pollution Prevention and Control (Scotland) Regulations 2000 (SSI 2000 No. 323).
- Pollution Prevention and Control (England and Wales) (Amendment) Regulations 2001 (SI 2001 No. 503).
- Pollution Prevention and Control (England and Wales) (Amendment) Regulations 2002 (SI 2002 No. 275) and Amendment No. 2, 2002 (SI 2002 No. 2197).

KEY DATES

- 2000 Regulations in force from 1/8/00 (Scotland 28/9/00).
- 2001 Regulations in force from 1/4/01.
- 2002 Regulations in force from 1/4/02 and 25/7/02.

PURPOSE

- To achieve a high level of protection of the environment taken as a whole by, in particular, preventing or, where that is not practicable, reducing emissions into the air, water and land.
- To introduce a more integrated approach to controlling pollution from industrial sources taking into account emissions into air, water or land; waste avoidance or minimisation; accident avoidance; minimisation of noise, heat and vibrations; and energy efficiency.

APPROACH

- PPCR 2000 sets out the main current framework for industrial and air pollution control, which provides that no person may operate an installation or mobile plant except under and to the extent authorised by a permit issued by the regulator.
- An installation is one where activities listed in Part 1 of Schedule 1 PPCR are carried out. The list provides Part A(1) activities subject to IPPC by the Environment Agency, Part A(2) activities subject to IPPC by local authorities and Part B activities subject to LAPPC by local authorities.
- Any permit granted will be subject to the following conditions:
  - All appropriate measures are taken against pollution in particular through application of Best Available Techniques (BAT) (see below).
  - No significant pollution is caused.
  - Waste production is avoided or recovered. Where this is not possible waste must be disposed of in a manner to produce least impact on the environment.
  - Energy is used efficiently.
  - Measures are taken to avoid accidents and limit their consequences.
  - Necessary measures are taken on closure of the installation to avoid any pollution risk and to return the site to a satisfactory condition.
- BAT involves the technology used and the way it is operated to ensure a high level of environmental protection as a whole by generally reducing emissions and environmental impact. It also takes into account the balance between the costs and environmental benefits.
- A basic requirement of permits is to impose emission limit values based on BAT.

REGULATORY AUTHORITY

- Environment Agency.
- Local Authority.
- Copies of applications for permits must be submitted by the Environment Agency or local authority to a variety of bodies, *e.g.*

Food Standards Agency, Health Authority in whose area instillation or mobile plant will operate.

COVERAGE

- Installations defined as "where one or more of the activities specified in Part 1 of Schedule 1 PPCR are carried out" (see *PPCR Regulated Installations* for more detail).
- For new installations and existing installations undergoing a substantial change (one that may have a negative effect on humans or the environment) IPPC permits are required from 31st October 1999.
- For existing Integrated Pollution Control (IPC) permits new IPPC permits are required from dates specified in PPCR and phased in on a sector basis by October 2007. 2002 Amendment Regulations defer by one year the periods for which IPPC applications need to be made to local authorities.

COMPLIANCE REQUIREMENTS

- Must apply for permit to operate an installation (whether existing or new) and the application must be made in accordance with PPCR.
- Must supply regulator with emission data to enable them to check compliance with permit.
- Allow regulator to undertake independent monitoring and inspections of the installation to check compliance.
- Must pay relevant fee or charge to regulatory authority.
- Must advertise the application in one or more local newspapers and in the London Gazette.
- Must operate within specified emission limits and any other conditions.
- Must notify any relevant change in operation or any transfer of installation to regulatory authority.
- Regulatory authority can serve: enforcement notice (where contravention occurs, it can specify steps and time for remedy); suspension notice (if there is an imminent risk of serious pollution). Regulatory authority can also revoke an authorisation.

DISCLOSURE

- Regulatory authority may request information from the applicant.
- Public register to be maintained by the regulatory authority, but it may not include information deemed to be commercially confidential or information affecting national security.

COMPLEMENTARY LEGISLATION

- Environmental Protection Act 1990 Part I (EPA).
- EPA 1990 Part III.
- Environmental Protection (Prescribed Processes and Substances) Regulations 1991 (as amended).
- Environmental Protection (Applications, Appeals and Registers) Regulations 1991 (as amended).
- Environmental Protection (Authorisation of Processes) (Determination Periods) Order 1991 (as amended).
- Clean Air Act 1993.
- Environment Act 1995.
- Environmental Protection (Prescribed Processes and Substances *etc.*) (Amendment) (Petrol Vapour Recovery) Regulations 1996.
- Landfill (England and Wales) Regulations 2002.

RELATED EC LEGISLATION

- EC Directive 84/360 on Industrial Plant Emissions.
- EC Directive 88/609 on Large Combustion Plants Emissions.
- EC Directive 96/61 on Integrated Pollution Prevention and Control.
- EC Directive 94/63 on the control of volatile organic compound emissions from petrol storage and distribution.
- EC Directive 99/31 on the Landfill of Waste.
- Plus the many other pollutant specific directives which are not accounted for in these summaries.

COMMENTS

- IPC and Air Pollution Control (APC) authorisations issued

under EPA are still in force until a new IPPC permit is required. Main differences are as follows:

- IPPC takes into account a wider range of environmental impacts.
- IPPC will apply to a wider range of industries.
- IPPC regulated industries are referred to as "installations" as opposed to "processes" under IPC. This gives a more integrated approach as a whole installation must be permitted rather than just individual processes within the installation.
- Both IPC and APC authorisations will require that the applicant apply the Best Available Techniques Not Entailing Excessive Costs (BATNEEC) so as to prevent/minimise polluting emissions whereas IPPC require BAT. In the case of IPC authorisations, if more than one BATNEEC process is available to the applicants then they must prove that they have selected the Best Practicable Environmental Option (BPEO) in pollution control and are not simply moving pollution.

- 1999 Act provides the framework for the details to be made by Regulation *e.g.* PPCR.
- Guidelines to establish which techniques are BAT are to be published by the European Commission's IPPC Bureau. These will be ready in the next couple of years. In the intervening period guidance that already exists from IPC and APC will continue to be used.
- Petrol Vapour Recovery Regulations added storage, loading and unloading of petrol at terminals and service stations to processes requiring APC authorisation under EPA. This is now included in PPCR Part B activity and therefore requires an LAPPC permit.
- PPCR specifies the procedure for granting permits and the information that must be provided in an application.
- 2002 Landfill Regulations disapplies the power to set conditions in PPCR for landfills as the Landfill Regulations provides alternative powers. The Environment Agency is to classify landfills as for hazardous waste, non-hazardous waste or inert waste. Conditions relating to waste acceptance, monitoring procedures, operational control, closure and after-care procedures, and general requirements are specified.

RELATED DOCUMENTATION

- Application for permit/variation and related correspondence.
- Documentation relating to publicity/advertisement requirements.
- Permit.
- Monitoring information.
- Any notification of change/transfer.
- Any enforcement/suspension notice or revocation details.

7.2 PPCR Regulated Installations

LEGISLATION

- Pollution Prevention and Control (England and Wales) Regulations 2000 (SI 2000 No. 1973).
- Pollution Prevention and Control (Scotland) Regulations 2000 (SSI 2000 No. 323).

KEY DATES

- In force from 1/8/00 (Scotland 28/9/00).

PURPOSE

- Define and list activities at an installation for which an IPPC permit is required.
- Designate period within which application to be made.

APPROACH

- Designate activities for which an IPPC permit is required in Part 1 of Schedules 1 PPCR. Part A(1) subject to IPPC by the Environment Agency, Part A(2) subject to IPPC by local authorities and Part B subject to LAPPC by local authorities.
- Sites operating any of these activities at an installation will require a permit to operate under PPCR.

REGULATORY AUTHORITY

- Environment Agency.
- Local Authority.

COVERAGE

- Part A and B activities are split into the following groups:
  1. Energy industries including combustion activities and gasification, liquefaction and refining activities.

2. Production and processing of metals including ferrous metals, non-ferrous metals, surface treating metals and plastic materials.
3. Mineral industries including production of cement and lime, activities involving asbestos, manufacturing glass and glass fibre, production of other mineral fibres, other mineral activities and ceramic production.
4. Chemical industries including organic chemicals, inorganic chemicals, chemical fertiliser production, plant health products and biocides, pharmaceutical production, explosives production, manufacturing activities involving carbon disulfide or ammonia and the storage of chemicals in bulk.
5. Waste management including disposal of waste by incineration, disposal of waste by landfill, disposal of waste other than by incineration or landfill, recovery of waste and the production of fuel from waste.
6. Other activities including paper, pulp and board manufacturing activities, carbon activities, tar and bitumen activities, coating activities, printing and textile treatments, manufacture of dyestuffs, printing ink and coating materials, timber activities, treatment of animal and vegetable matter, and food industries.

COMPLIANCE REQUIREMENTS

- Where Regulations indicate activity is to be carried out at an installation one must apply for IPPC permit within specified time limits.

DISCLOSURE

- Regulatory authority may request information from the applicant.
- Public register to be maintained by the regulatory authority, but it may not include information deemed to be commercially confidential or information affecting national security.

COMPLEMENTARY LEGISLATION

- EPA 1990 Part I.
- Environmental Protection (Prescribed Processes and Substances) Regulations 1991.
- Environmental Protection (Amendment of Regulations) Regulations 1991.
- Environmental Protection (Prescribed Processes and Substances) (Amendment) Regulations 1992, (Amendment) Regulations 1993 & (Amendment No.2) Regulations 1993, (Amendment) Regulations 1994 & (Amendment No.2) Regulations 1994, (Amendment) Regulations 1995.
- Pollution Prevention and Control Act 1999.
- Landfill (England and Wales) Regulations 2002.
- Pollution Prevention and Control (Designation of Council Directives on Large Combustion Plants, Incineration of Waste and National Emission Ceilings) Order 2002.

RELATED EC LEGISLATION

- EC Directive 84/360 on Industrial Plant Emissions.
- EC Directive 99/31 on the Landfill of Waste.
- EC Directive 2000/76 on the incineration of waste.
- EC Directive 2001/80 on the limitation of emissions of certain pollutants into the air from large combustion plants.
- EC Directive 2001/81 on national emission ceilings for certain atmospheric pollutants.

COMMENTS

- Certain activities are defined by the release of specified substances into an environmental media *e.g.* formulating pharmaceutical products if that may result in the release of a specified substance into water. Schedule 1 Part 2 PPCR lists specified substances for release into air and water at quantities greater than background levels. The list for air includes oxides of sulphur and carbon, halogens and their compounds. The list for water includes mercury and its compounds, all isomers of DDT. Therefore, if an installation releases substances into the appropriate media and comes within the installation definition then an IPPC permit will be required.

- Landfill Regulations 2002 makes all landfills covered by these Regulations Part A(1) installations for the purposes of PPCR.
- 2002 Order in force from 9th October 2002 provides that the Directives concerning incineration of waste, large combustion plants and national emission ceilings are relevant. Directives in determining provisions for regulating polluting activities *i.e.* these Directives can be taken into account in deciding what pollution control measures are required.

RELATED DOCUMENTATION

- Application for permit/variation and related correspondence.
- Documentation relating to publicity/advertisement requirements.
- Permit.
- Monitoring information.
- Any notification of change/transfer.
- Any enforcement/suspension notice or revocation details.

8 AIR POLLUTION

8.1 Air Quality Standards

LEGISLATION

- Air Quality Standards Regulations 1989 (SI 1989 No. 317).
- Air Quality Standards (Amendment) Regulations 1995 (SI 1995 No. 3146).
- Air Quality Limit Values Regulations 2001 (SI 2001 No. 2315).
- Air Quality Limit Values (Scotland) Regulations 2001 (SSI 2001 No. 224).
- Air Quality Limit Values (Wales) Regulations 2001 (SI 2001 No. 2683).

KEY DATES

- In force from 31/3/89.
- 1995 Regulations in force from 31/12/95.
- 2001 Regulations in force from 19/7/01.

PURPOSE

- To ensure acceptable quality of ambient air.
- Implement EC Directive 96/62 on ambient air quality assessment and management and EC Directive 99/30 relating to limit values for sulphur dioxide, nitrogen dioxide, and oxides of nitrogen, particulate matter and lead in ambient air.
- Implement EC Directives on Air Quality Limit Values for Sulphur Dioxide and Suspended Particulates (80/779), Lead (82/884) and Nitrogen Dioxide (85/203).

APPROACH

- Impose quality standards or Air Quality Limit Values (AQLVs) for maximum concentrations of sulphur dioxide and suspended particulates, lead and nitrogen dioxide in ambient air.

- Secretary of State must identify, assess (using specified methods) and classify each zone. This classification must be reviewed every five years or if significant changes.
- Secretary of State under a duty to take necessary measures (taking an integrated approach to the protection of air, water and soil) to ensure these limit values are not exceeded in each zone. The limit values and dates to achieve them are set out in Schedule 1 to the Regulations.
- If limit values are exceeded the Secretary of State must draw up a plan for improvement.
- Sets up monitoring and sampling stations to supply the data necessary for the fulfilment of the requirements of the Directives.

REGULATORY AUTHORITY

- Secretary of State for the Environment, Food and Rural Affairs.

COVERAGE

COMPLIANCE REQUIREMENTS

DISCLOSURE

COMPLEMENTARY LEGISLATION

- Environmental Protection Act 1990.
- Pollution Prevention and Control (England and Wales) Regulations 2000.
- Pollution Prevention and Control (Scotland) Regulations 2000.

RELATED EC LEGISLATION

- EC Directive 80/779 on Air Quality Limit Values and Guidelines for Sulphur Dioxide and Suspended Particulates.
- EC Directive 82/884 on a Limit Value for Lead in the Air.

- EC Directive 85/203 on Air Quality Standards for Nitrogen Dioxide
- EC Directive 96/62 on ambient air quality assessment and management.
- EC Directive 99/30 relating to limit values for sulphur dioxide, nitrogen dioxide, and oxides of nitrogen, particulate matter and lead in ambient air.

COMMENTS

- EC Directives 80/779, 82/884 and 85/203 have all been repealed by EC Directive 99/30 but are still included as transitional provisions last up to 2005 and 2010.
- Industrial emissions are controlled either:
    - By local authorities by means of "local air pollution control" and by the Environment Agency under "integrated pollution control" under Part I of the Environment Protection Act 1990.
    - By local authorities and the Environment Agency using their powers under the Pollution Prevention and Control (England and Wales) Regulations 2000.

RELATED DOCUMENTATION

8.2 Air Quality Standards and Objectives

LEGISLATION

- Environment Act 1995 Part IV.
- Air Quality (England) Regulations 2000 (SI 2000 No. 928).
- Air Quality (Wales) Regulations 2000 (SI 2000 No. 1940).
- Air Quality (Scotland) Regulations 2000 (SSI 2000 No. 97).

KEY DATES

- In force from 23/12/97.
- 2000 England Regulations in force from 6/4/00.
- 2000 Wales Regulations in force from 1/8/00.
- 2000 Scotland Regulations in force from 7/4/00.

PURPOSE

- The 1995 Act introduced a new system of local air quality management to ensure that the air quality standards and objectives for key pollutants will be achieved.
- These standards are now set out in the 2000 Regulations, and are the same as the *Air Quality Strategy for England, Scotland, Wales and Northern Ireland* (published January 2000).

APPROACH

- Part IV of the 1995 Act places all local authorities under a duty to carry out regular reviews of current and likely future air quality in their areas. These reviews have to be accompanied by an assessment of whether any prescribed air quality standards or objectives are being achieved or are likely to be achieved within the relevant period.
- The objectives for the seven pollutants concerned, which are to be achieved by the relevant period, are set out by the 2000 Regulations as follows:

Substance	*Air quality objective levels*	*Air quality objective dates*
Benzene	16.25 micrograms per cubic metre or less, when expressed as running annual mean	(31/12/03)
1,3-Butadiene	2.25 micrograms per cubic metre or less, when expressed as running annual mean	(31/12/03)
Carbon monoxide	11.6 milligrams per cubic metre or less, when expressed as running 8 hour mean	(31/12/03)
Lead	0.5 micrograms per cubic metre or less, when expressed as an annual mean	(31/12/04)
	0.25 micrograms per cubic metre or less, when expressed as an annual mean	(31/12/08)
Nitrogen dioxide	200 micrograms per cubic metre, when expressed as an hourly mean, not to be exceeded more than 18 times a year	(31/12/05)
	40 micrograms per cubic metre or less, when expressed as an annual mean	(31/12/05)
PM10	50 micrograms per cubic metre or less, when expressed as a 24 hour mean, not to be exceeded more than 35 times a year	(31/12/04)
	40 micrograms per cubic metre or less, when expressed as an annual mean	(31/12/04)
Sulphur dioxide	125 micrograms per cubic metre or less, when expressed as a 24 hour mean, not to be exceeded more than 3 times a year	(31/12/04)
	350 micrograms per cubic metre or less, when expressed as an hourly mean, not to be exceeded more than 24 times a year	(31/12/04)
	266 micrograms per cubic metre or less, when expressed as a 15 minute mean, not to be exceeded more than 35 times a year	(31/12/05)

- If the local authorities assess that the air quality is likely to fall short of the national objectives within the relevant period, the local authority concerned must declare an "air quality management area" (AQMA) and set up an action plan to achieve compliance.

REGULATORY AUTHORITY

- Local authority.
- Secretary of State.

COVERAGE

COMPLIANCE REQUIREMENTS

DISCLOSURE

COMPLEMENTARY LEGISLATION

RELATED EC LEGISLATION

- EC Directive 96/92 on ambient air quality assessment and management.

COMMENTS

RELATED DOCUMENTATION

8.3 Asbestos

LEGISLATION

- Control of Asbestos in the Air Regulations 1990 (SI 1990 No. 556).

KEY DATES

- In force partly on 5/5/90; fully on 30/6/91.

PURPOSE

- Prevent and reduce environmental pollution by asbestos.
- As originally enacted, implemented as regards emissions into the air EC Directive 87/217 on asbestos (see Comments).

APPROACH

- Impose general requirement that asbestos fibres or dust emitted into the air resulting from the workings of asbestos products, or from the demolition of buildings/installations containing asbestos and the removal from them of asbestos, do not cause "significant environmental pollution" (Regulation 4).

REGULATORY AUTHORITY

- Secretary of State.

COVERAGE

- All persons undertaking activities involving:
  - The working of products containing asbestos or
  - The demolition of buildings/installations containing asbestos and the removal from them of asbestos.

COMPLIANCE REQUIREMENTS

- All persons undertaking activities involving the working of products containing asbestos must ensure that "significant environmental pollution" is not caused by the emission into the air of asbestos fibres or dust resulting from the workings of asbestos products.

- All persons undertaking activities involving the demolition of buildings/installations containing asbestos and the removal from them of asbestos must ensure that "significant environmental pollution" is not caused.

DISCLOSURE

COMPLEMENTARY LEGISLATION

- Health and Safety at Work *etc.* Act 1974.
- Asbestos (Licensing) Regulations 1983.
- Environmental Protection Act 1990 (EPA).
- Asbestos (Prohibitions) Regulations 1992.
- Control of Asbestos at Work Regulations 2002.

RELATED EC LEGISLATION

- EC Directive 87/217 on the prevention and reduction of pollution by asbestos.

COMMENTS

- Breach of the Regulation's requirements is a criminal offence.
- Note: Former provisions (1) prescribing a limit value for the discharge of asbestos from outlets into the air at all plants involved with the "use of asbestos" and (2) requiring the regular measurement of asbestos emissions from such plants at least every six months, in line with EC Directive 87/217 on asbestos, have now lapsed after the EPA 1990 has repealed relevant sections of the Health and Safety at Work *etc.* Act 1974.

RELATED DOCUMENTATION

8.4 Clean Air

LEGISLATION

- Clean Air Act 1993.
- Dark Smoke (Permitted Periods) Regulations 1958 (SI 1958 No. 498).
- Clean Air (Arrestment Plant) (Exemption) Regulations 1969 (SI 1969 No. 1262).
- Clean Air (Height of Chimneys) (Exemption) Regulations 1969 (SI 1969 No. 411).
- Clean Air (Emission of Grit and Dust from Furnaces) Regulations 1971 (SI 1971 No. 162).

KEY DATES

- In force from 27/8/93.

PURPOSE

- Regulate air pollution by restricting or prohibiting certain emissions into atmosphere, and in particular to control the emission of smoke and the emission of grit and dust.

APPROACH

- Prohibits emissions of "dark smoke" from industrial/trade premises and from chimneys of any building or furnace serving boilers and industrial plant (subject to exceptions contained in the Dark Smoke (Permitted Periods) Regulations 1958 and some general defences to prosecution, *e.g.* that the emission was due to the failure of a furnace).
- Establishes requirement of prior approval by local authorities for the installation of a furnace in any building, or in any fixed boiler or industrial plant, and no such installation may be made unless the furnace is so far as practicable capable of being operated continuously without emitting smoke when burning fuel of a type for which the furnace was designed. The provisions do not apply to domestic furnaces.

- Details of the permitted amounts of grit and dust to be emitted from furnaces are contained in the Clean Air (Emission of Grit and Dust from Furnaces) Regulations 1971. The 1993 Act also requires that grit and dust arrestment plant be fitted to new industrial furnaces which burn certain types of fuel, or which operate at certain capacities (subject to exemptions contained in Clean Air (Arrestment Plant) (Exemption) Regulations 1969), and the type of arrestment plant must also be approved by the local authority.
- Establishes requirement of prior approval by local authorities of furnace chimney heights before furnace is used to burn pulverised fuel or other fuels at certain rates (subject to exemptions contained in Clean Air (Height of Chimneys) (Exemption) Regulations 1969).
- Provides local authorities with powers to establish smoke control areas where smoke emissions can be prohibited.
- Provides Secretary of State with powers to regulate composition of any kind of motor fuel and sulphur content of gas oil for the purpose of reducing air pollution (see Comments).

REGULATORY AUTHORITY

- Local authority.
- Secretary of State.

COVERAGE

- "Dark smoke" is smoke darker than shade 2 on Ringelmann Chart.
- Provisions concerned with arrestment plant and chimney heights apply to furnaces which:
  - Burn pulverised fuel.
  - Burn any other solid matter at a rate of 45.4 kilograms or more an hour.
  - Burn any liquid or gaseous matter at a rate equivalent to 366.4 kilowatts per hour.
- Smoke control areas are districts declared as such by regulatory authority under smoke control order.
- Note: Provisions of Act do not apply to installations covered under Pollution Prevention and Control Regulations 2000.

COMPLIANCE REQUIREMENTS

- Must comply with restrictions on emission of dark smoke from chimneys and industrial/trade premises. (Details of the restrictions are determined by the Dark Smoke (Permitted Periods) Regulations 1958. Two separate circumstances are considered: continuous and intermittent emissions. The limit for continuous emission of dark smoke, other than in the course of soot-blowing, is four minutes. The determination of intermittent emissions is based on aggregate emissions for a period of eight hours).
- Must obtain prior approval from local authority for installation of any new non-domestic furnace in a building or any fixed boiler/industrial plant, and must ensure that the furnace is so far as practicable capable of being operated continuously without emitting smoke when burning fuel of type for which the furnace was designed.
- Must comply with any limits for the emission of grit and dust from industrial furnaces contained in Clean Air (Emission of Grit and Dust from Furnaces) Regulations 1971. Where no limits have been prescribed, the occupier of building in which furnace is operated is placed under general duty to minimise grit and dust emissions.
- Must also fit grit and dust arrestment plant to industrial furnaces which burn certain types of fuel or operate at certain capacities (unless exempt), and details of the arrestment plant must also be given to the local authority. Must apply to local authority for any exemption.
- Must obtain prior approval of furnace's chimney height from local authority before furnace is used to burn pulverised fuel or other fuels at certain rates (unless exempt).
- Must comply with prohibitions/use authorised fuel in smoke control area.
- Occupier served with information notice by local authority must supply specified information on emissions within 6 weeks (Note: grounds of appeal exist).

DISCLOSURE

- Local Authority may issue information notice requiring occupier of any premises to furnish information on emissions.

COMPLEMENTARY LEGISLATION

- Clean Air (Measurement of Grit and Dust from Furnaces) Regulations 1971.
- The Motor Fuel (Composition and Content) Regulations 1994.
- Environmental Protection Act 1990 Part I.
- Environment Act 1995.
- Pollution Prevention and Control (England and Wales) Regulations 2000.
- Pollution Prevention and Control (Scotland) Regulations 2000.
- The Sulphur Content of Liquid Fuels (Scotland) Regulations 2000.
- The Sulphur Content of Liquid Fuels (England and Wales) Regulations 2000.

RELATED EC LEGISLATION

COMMENTS

- The Sulphur Content of Liquid Fuels (England and Wales) Regulations 2000 prohibits the use of gas oil with a sulfur content of more than 0.2% by weight.

RELATED DOCUMENTATION

- Prior approval for furnaces.
- Any exemption sought for arrestment plant.
- Monitoring/measuring information.
- Prior approval for chimney heights.
- Any information notice served.

8.5 Climate Change Levy

LEGISLATION

- Finance Act 2000 s 30 and Schedule 6.
- Climate Change Levy (Registration and Miscellaneous Provisions) Regulations 2001 (SI 2001 No. 7).
- Climate Change Levy (Combined Heat and Power Stations) Exemption Certificate Regulations 2001 (SI 2001 No. 486).
- Climate Change Levy (General) Regulations 2001 (SI 2001 No. 838).
- Climate Change Levy (General) (Amendment) Regulations 2002 (SI 2002 No. 1152).

KEY DATES

- Levy in force from 1/4/01.
- SI 2001 No. 7 in force from 29/1/01.
- SI 2001 No. 486 in force from 19/3/01.
- SI 2001 No. 838 in force from 1/4/01.
- 2002 Regulations in force from 1/6/02.

PURPOSE

- Introduce climate change levy on supplies for industrial and commercial purposes of energy that produce carbon dioxide emissions (electricity, gas *etc.*), ostensibly to help UK meet its obligations under UN Framework Convention on Climate Change and Kyoto Protocol.

APPROACH

- Finance Act 2000 Schedule 6 introduces climate change levy on supplies for industrial and commercial purposes of energy in the form of electricity, gas, petroleum gas and other gaseous hydrocarbons supplied in a liquid state, plus other taxable commodities such as coal.
- Person liable to account for the levy is the supplier (with some exceptions).

- Rates of levy are set at: 0.43 p per kWh for electricity; 0.15 p per kWh for gas; 0.96 p per kg for liquid petroleum gas *etc.* and 1.17p per kg for coal and any other taxable commodity.
- Levy applies to supplies for industry and commerce, but not domestic or charity use, fuel for transport, energy used to produce another energy product and exemptions also apply to electricity generated by renewable sources (see Coverage).
- Levy is charged at a reduced rate of 20% where suppliers in the energy-intensive sector have concluded approved climate-change agreements.

REGULATORY AUTHORITY

- Customs and Excise.

COVERAGE

- Levy applies to industry (including fuel industries), commerce, agriculture, public administration and other services.
- Does NOT apply to domestic or charity use, fuel for transport, energy used to produce another energy product (*e.g.* coal used to generate electricity) and use to manufacture feedstocks *etc.*
- Exemptions also apply to electricity generated by renewable sources such as wind and solar power, and by good quality combined heat and power plants (which are either fully or partly exempt from levy).

COMPLIANCE REQUIREMENTS

- Various requirements for registration by suppliers with Customs and Excise, in line with 2001 Regulations and Part V of Schedule 6 to Finance Act 2000.
- Pay the levy if appropriate.

DISCLOSURE

COMPLEMENTARY LEGISLATION

- Climate Change Agreements (Eligible Facilities) Regulations 2001 (SI 2001 No. 662).
- Climate Change Levy (Electricity and Gas) Regulations 2001 (SI 2001 No. 1136).
- Climate Change Levy (Solid Fuel) Regulations 2001 (SI 2001 No. 1137).
- Climate Change Levy (Use as Fuel) Regulations 2001 (SI 2001 No. 1138).
- Climate Change Agreements (Energy-intensive Installations) Regulations 2001 (SI 2001 No. 1139).
- Climate Change Levy (Combined Heat and Power Stations) Prescribed Conditions and Efficiency Percentages Regulations 2001 (SI 2001 No. 1140).

RELATED EC LEGISLATION

COMMENTS

- SI 2001, No. 486 relates to exemption certificates issued by the Secretary of State for combined heat and power plants. Provision is made relating to applications for, and the content and validity of, certificates. Also provided are requirements to vary the certificate where there has been a material change in the circumstances of the combined heat and power station. This could include upgrading the levy from partial to full exemption. SI 2001, No. 1140 provides two conditions before a combined heat and power plant is entitled to full exemption. The first is that the station is assessed against the Combined Heat and Power Quality Assurance Standard while the second relates to the station's efficiency.
- Provision is made for climate change agreements where certain conditions are satisfied including agreement with Secretary of State, reduction in climate change levy and targets for improvement are set and reviewed every five years. By SI 2001 662, in force from 1st April 2001, provides that a facility is only

eligible for inclusion in a climate change agreement where it is likely that at least 90% of the energy supplied to the facility will be used within an energy intensive installation.

- SI 2001 No. 838 makes provision for the mechanics of payment of climate change levy and how to calculate the amount due. It provides for reduction of payment in certain circumstances *e.g.* to avoid double payment. Certain requirements are echoed from older provisions in force from other taxes *e.g.* representation in case of death, incapacity or insolvency and the transfer of a business as a going concern.
- 2002 Regulations make provision for an annual accounting scheme – registered traders can account for and pay climate change levy on an annual basis provided they satisfy certain specified criteria.

RELATED DOCUMENTATION

8.6 Ozone-depleting Substances

LEGISLATION

- EC Regulation on Substances that Deplete the Ozone Layer (2037/2000/EC) (EC Regulation).
- EC Regulation on substances that deplete the ozone layer, as regards metered dose inhalers and medical drug pumps (2038/2000/EC).
- EC Regulation on substances that deplete the ozone layer, as regards the base year for the allocation of quotas of hydrochlorofluorocarbons (2039/2000/EC).
- Environmental Protection (Controls on Ozone-depleting Substances) Regulations 2002 (SI 2002 No. 528).

KEY DATES

- 2000 Regulations in force in Member States from 30/9/00.
- 2002 Regulations in force from 31/3/02.

PURPOSE

- To prevent damage to the ozone layer.
- The EC Regulation implements the requirements of the Montreal Protocol on Substances that Deplete the Ozone Layer (and subsequent amendments) and the Vienna Convention for the Protection of the Ozone layer (to which the EC and all of its Member States, including the UK, are party). *N.B.* It repeals and replaces all the earlier implementing EC Regulations.

APPROACH

- Prohibits or restricts the production, placing on the market, import, export, and use of various ozone-depleting substances beyond a variety of dates. (*N.B.* The EC Regulation is directly applicable law in the UK. Therefore its requirements bind producers, suppliers, importers, exporters and users of the various substances directly (although Member States are required to deal with a number of ancillary matters by means of their domestic legislation)).

- Provides requirements for recovery and potential leakage of controlled substances.
- In terms of sanctions, it is up to each Member State to determine the penalties to be imposed in the event of any failure to comply with the EC Regulation. These are set out for the UK in SI 2002 No. 528 which also confer various enforcement powers on authorised persons and officers.

REGULATORY AUTHORITY

- European Commission.
- The Secretary of State is the competent authority for the purposes of the EC Regulation.

COVERAGE

- "Controlled substances" covered by the Regulation include chlorofluorocarbons (CFCs), other fully halogenated CFCs, halons, carbon tetrachloride, 1,1,1-trichloroethane, methyl bromide, hydrobromofluorocarbons and hydrochlorofluorocarbons.

COMPLIANCE REQUIREMENTS

- Must comply with the prohibition on the production, placing on the market or use of CFCs, other fully halogenated CFCs, halons, carbon tetrachloride, 1,1,1-trichloroethane, hydrobromofluorocarbons (subject to exceptions).
- Producers must ensure that production level of methyl bromide – or level of methyl bromide which they place on the market or use for their own account – from 1 January to 31st December 1999 and in each 12-month period thereafter does not exceed 75% of the level which they produced in 1991. The 75% requirement is reduced to 40% for 1 January to 31st December 2002 and to 25% for the calendar years 2003 and 2004. Production of methyl bromide is totally banned after 31st December 2004.
- Similar staged reduction in production of hydrochlorofluorocarbons applies culminating in no production after 31st December 2025.

- Must obtain import licence from Commission, to import a controlled substance from a non EC country, to carry out release for free circulation in the Community or inward processing of controlled substances. Certain controlled substances specified in Annex 1 to EC Regulation shall not be imported for inward processing. Quantitative limits for import from non EC Countries are provided.
- Must comply with prohibition on the release for free circulation in the Community of controlled products from any State not party to the Montreal Protocol.
- Export of controlled substances or products containing those substances is prohibited with specified exceptions. Such exceptions shall be subject to authorisation.
- Must comply with restrictions on the recovery of used controlled substances.
- Must take all precautionary measures practicable to prevent leakages of controlled substances, *e.g.* fixed equipment with a refrigerating fluid charge of more than 3 kg shall be checked for leakages annually. Must also take all precautionary measures practicable to prevent and minimise leakages of methyl bromide from fumigation installations and operations in which methyl bromide is used.
- Each producer, importer and exporter of controlled substances must – each year before March 31st – communicate to the Commission (also sending a copy to the competent authority of the Member State concerned) various details concerning production, import or export of each controlled substance in respect of the previous calendar year.

DISCLOSURE

- Yearly communication to Commission.

COMPLEMENTARY LEGISLATION

RELATED EC LEGISLATION

COMMENTS

- EC Regulation 2038/2000 provided that export of metered dose inhalers and medical drug pumps containing chlorofluorocarbons to developing countries should not be restricted provided use is permitted in the Community market.
- EC Regulation 2039/2000 changed the base year for allocation of quotas of hydrochlorofluorocarbons from 1996 to 1999.

RELATED DOCUMENTATION

- Data on production/marketing/use levels of methyl bromide (back to 1991).
- Any import licence.
- Details concerning production, import or export of each controlled substance in respect of previous calendar year.

8.7 Statutory Nuisances

LEGISLATION

- Environmental Protection Act 1990 Part III: Statutory Nuisances and Clean Air.
- Statutory Nuisance (Appeals) Regulations 1995 (SI 1995 No. 2644).

KEY DATES

- In force from 1/1/91.

PURPOSE

- Protect human health and prevent/minimise pollution in the environment by giving local authorities preventative and remedial powers to deal with various types of pollution from factories, shops and other premises.

APPROACH

- Requires local authorities to inspect their area periodically to detect whether any statutory nuisance (as listed in the Act – see Coverage) exists.
- Where local authority is satisfied that a statutory nuisance exists, or is likely to occur, it must serve an abatement notice on the responsible person (or the owner of the premises where that person cannot be found or the nuisance has not yet occurred). The abatement notice can prohibit or restrict the occurrence or re-occurrence of the nuisance and require the execution of any necessary work.
- A defence is available in most cases that the industry, trade or business is using Best Practicable Means (BPM) to prevent, or counteract the effects of, the nuisance taking into account local conditions, current technical knowledge and financial implications. BPM can also cover the design, construction and maintenance of buildings and enclosures.

REGULATORY AUTHORITY

- Local authority.

COVERAGE

- Regulations apply to any premises, including industrial/trade/business premises, where statutory nuisance exists.
- List of statutory nuisances includes any:
  - Smoke, fumes, gases, dust, steam, smell or other effluvia arising on industrial/trade/business premises, any accumulation or deposit, or noise.
  - Activity which is "prejudicial to health or a nuisance".
- Nuisance is not defined by Act and is determined by common law in the courts.

COMPLIANCE REQUIREMENTS

- Where abatement notice has been served by the local authority, the responsible person (or the owner of the premises where that person cannot be found or the nuisance has not yet occurred) must comply with terms of notice, which can:
  - Prohibit or restrict the occurrence or reoccurrence of the nuisance.
  - Require the execution of any necessary work to abate the nuisance.
- Non-compliance with the terms of the abatement notice is a criminal offence. (See Comments).
- Local authority can recover expenses reasonably incurred in abating statutory nuisance from responsible person.

DISCLOSURE

COMPLEMENTARY LEGISLATION

- EPA 1990 Part I, Part IIA.
- Clean Air Act 1993.

- Noise and Statutory Nuisance Act 1993.
- Environment Act 1995.
- Pollution Prevention and Control (England and Wales) Regulations 2000.

RELATED EC LEGISLATION

COMMENTS

- Grounds of Appeal exist. See further the Statutory Nuisance (Appeals) Regulations 1995. Persons served with the abatement notice may appeal against it to a magistrates court within 21 days.

RELATED DOCUMENTATION

- Any abatement notice served.
- Any charging notice.

8.8 Sulphur Content of Liquid Fuels

LEGISLATION

- Sulphur Content of Liquid Fuels (England and Wales) Regulations 2000 (SI 2000 No. 1460).
- Sulphur Content of Liquid Fuels (Scotland) Regulations 2000 (SSI 2000 No. 169).

KEY DATES

- In force from 27/6/00 (Scotland 30/6/00).

PURPOSE

- Reduce air pollution by regulating the sulphur content of liquid fuels.
- Implement provisions of EC Directive 99/32 relating to the sulphur content of certain liquid fuels.

APPROACH

- Prohibit the use of certain oils with specified sulphur contents.
- Require sampling to check the sulphur content of certain oils.

REGULATORY AUTHORITY

- Local Authority.

COVERAGE

- Gas oils as defined but excluding diesel fuels and fuels used in non-road mobile machinery and agricultural tractors.
- Heavy fuel oil and marine gas oil as defined.
- Excludes fuel intended for processing prior to final combustion and the use of fuel to be processed in the refining industry.

COMPLIANCE REQUIREMENTS

- Must not use heavy fuel oil on or after 1st January 2003 with a sulphur content exceeding 1% except in a combustion plant which is subject to a permit with conditions relating to sulphur dioxide emissions.
- Must not use gas oil or marine gas oil on or after 1st July 2000 with a sulphur content exceeding 0.2% by mass.
- Must not use gas oil or marine gas oil on or after 1st January 2008 with a sulphur content exceeding 0.1% by mass.

DISCLOSURE

COMPLEMENTARY LEGISLATION

- Clean Air Act 1993.
- Motor Fuel (Composition and Content) Regulations 1994.

RELATED EC LEGISLATION

- EC Directive 99/32 relating to the sulphur content of certain liquid fuels.

COMMENTS

- Provision is made for sampling of fuels, using approved specified methods, to ensure compliance.
- Powers of entry contained in the Environment Act 1995 apply for the purposes of these Regulations.

RELATED DOCUMENTATION

- Combustion plant permits and associated correspondence.
- Sampling, analysis results.

9 GENETICALLY MODIFIED ORGANISMS

9.1 Deliberate Release and Marketing

LEGISLATION

- Environmental Protection Act Part VI Genetically Modified Organisms.
- Genetically Modified Organisms (Deliberate Release) Regulations 2002 (SI 2002 No. 2443).

KEY DATES

- In force from 17/10/02.

PURPOSE

- To ensure all appropriate measures are taken to avoid damage to the environment which may arise from the escape or release from human control of genetically modified organisms (GMOs).

APPROACH

- Imposes a requirement to obtain consent for deliberate release of GMOs.
- Consent also required for marketing of GMOs.

REGULATORY AUTHORITY

- Secretary of State for the Environment, Food and Rural Affairs

COVERAGE

- England and Wales only.

- Organism is widely defined in the Environmental Protection Act 1990 (EPA).
- An organism is 'genetically modified' if any of the genes or other genetic material in the organism:
  - Have been artificially modified, or
  - Are inherited or otherwise derived, through any number of replications, from genes or other genetic material (from any source) which were so modified.
- Higher plant means a plant belonging to the taxonomic group *Spermatophytae (Gymnospermae* or *Angiospermae).*
- Exemptions for marketing consent requirement for GMOs include an approved product which is marketed for a use for which it has approval and GMOs used exclusively for activities covered by the Genetically Modified Organisms (Contained Use) Regulations 2000.

COMPLIANCE REQUIREMENTS

- Must obtain a consent for the release of GMOs in all cases and circumstances in which GMOs are intended to be released.
- Must obtain a consent for the marketing of GMOs in all cases and circumstances in relation to the marketing of GMOs.
- Consent application:
  - Must be made in writing to the Secretary of State.
  - Must contain specified information:
    - For higher plants under the headings general information (*e.g.* title of the project); parental or recipient plant information; genetic modification information; genetically modified plant information; site release information; release information; control, monitoring, post-release and waste treatment plans; methodology information.
    - For any other case under the headings general information; organism information; vector characteristics; modified organisms characteristics; final form GMOs characteristics; release information; interactions between organisms and the environment; monitoring, control, waste treatment and emergency response plans; methodology information.

- For deliberate release consent must publish stated information in a national newspaper specified by the Secretary of State and send him/her a copy.
- Transitional provisions exist for consent where application is made before 17th October 2002 and the application has not yet been determined. In such cases the applicant has until 17th January 2003 to submit further information if required to comply with these Regulations.
- Renewal applications to market GMOs must be made before 17th October 2006 where consent issued before 17th October 2002 and no later than nine months before the expiry of the consent in all other cases.
- Procedural matters are specified in the Regulations *e.g.* Secretary of State duties and decision matters.
- General provisions are specified that must be in the consent *e.g.* period of validity of consent and scope

DISCLOSURE

- Consent application to Secretary of State.
- Register of consents.

COMPLEMENTARY LEGISLATION

- Genetically Modified Organisms (Contained Use) Regulations 2002.

RELATED EC LEGISLATION

- EC Directive 90/219 on the contained use of genetically modified microorganisms as amended by EC Directives 94/51 and 98/81.
- EC Directive 2001/18 on the deliberate release into the environment of genetically modified organisms.

COMMENTS

- An assessment report is prepared by the Secretary of State and forwarded to the European Commission. Information to be included in the assessment report is stated in the Regulations.

- EPA contains wide powers for inspectors to have a right of entry, inspection and to obtain information. This also includes powers to deal with imminent danger of damage to the environment.
- Exemptions exist from inclusion on the register of consents *e.g.* information would be contrary to the interests of national security.
- EPA and Regulations together provide the law for GMOs deliberate release and marketing and therefore must be read together to determine the legal requirements.

RELATED DOCUMENTATION

- Consent application.
- Genetic manipulations procedures.
- GMOs information.

10 WATER POLLUTION

10.1 Abandoned Mines

LEGISLATION

- The Mines (Notice of Abandonment) Regulations 1998 (SI 1998 No. 892).
- Water Resources Act 1991 Section 91B.

KEY DATES

- The 1998 Regulations and the new Section 91B in the Water Resources Act 1991 – which was inserted by the Environment Act 1995 – both came into force on 1/7/98.

PURPOSE

- Prevent pollution of controlled waters by abandoned mine waters.

APPROACH

- Section 91B requires mine operators to give at least six months' notice before abandoning a mine or part of a mine and let the Environment Agency know of any proposals to prevent or treat water pollution.
- The 1998 Regulations prescribe the information that must be contained in notices given under section 91B and the particulars which must be published by the operator in at least one local newspaper.
- The amendments made by the 1995 Act also remove after 31st December 1999 the statutory exemption against prosecution for the owners and former operators of mines who "knowingly permit" pollution of controlled waters by abandoned mine waters.

REGULATORY AUTHORITY

- Environment Agency.

COVERAGE

COMPLIANCE REQUIREMENTS

Mine operators are required to:

- Give the Environment Agency at least six months' notice of any proposed abandonment of a mine or part of a mine, on or after January 1st 1999.
- Include in the notice to the Environment Agency the operator's opinion of the abandonment's consequences and let them know of any proposals to prevent or treat water pollution.
- Publish in at least one local newspaper various details of the proposed abandonment.

DISCLOSURE

- Mine operators must publish in at least one local newspaper various details of the proposed abandonment.

COMPLEMENTARY LEGISLATION

- Environment Act 1995.
- Control of Pollution Act 1974 Section 30Z(1) and the Mines (Notification of Abandonment) (Scotland) Regulations 1998 impose similar requirements in Scotland.

RELATED EC LEGISLATION

COMMENTS

RELATED DOCUMENTATION

- Notice to Environment Agency.
- Advertisement in local newspaper.

10.2 Discharges to Controlled Waters

LEGISLATION

- Water Resources Act 1991 (WRA).
- Control of Pollution (Applications, Appeals and Registers) Regulations 1996 (SI 1996 No. 2971).
- Anti-pollution Works Regulations 1999 (SI 1999 No. 1006).

KEY DATES

- In force from 1/12/91.
- The 1996 Regulations came into force on 31/10/96.
- The 1999 Regulations came into force on 29/4/99.

PURPOSE

- Prevent/minimise pollution of controlled waters, including both surface and groundwater (see Coverage).

APPROACH

- The key provision of the Water Resources Act 1991 relating to protection of the aquatic environment is Section 85, under which a person commits an offence if he causes or knowingly permits any poisonous, noxious or polluting matter, including trade effluent or sewage effluent or any solid waste matter to enter any stream or controlled waters.
- It is a defence if release is made in accordance with a consent issued by Environment Agency under this Act, or with an authorisation granted under the Pollution Prevention and Control Act 1999 where release is from an IPPC regulated process.
- Environment Agency is under a duty to exercise its powers, including the setting and variation of consents, so as to achieve any statutory water quality objectives set under the WRA 1991 at all times, so far as it is practicable to do so.

REGULATORY AUTHORITY

- Environment Agency.
- Secretary of State.

COVERAGE

- "Controlled waters" include (see section 104):
  - Relevant territorial waters (*i.e.* the sea extending three miles out from specified baselines, despite the extension of the territorial limit to 12 miles).
  - Coastal waters (*i.e.* those extending landward from the same specified baselines to high tide/freshwater limits).
  - Inland freshwaters (*i.e.* Rivers, streams, canals, lakes and reservoirs).
  - Groundwaters (those contained in underground strata, bores or wellholes).

COMPLIANCE REQUIREMENTS

- Must apply for consent to discharge trade effluent/waste to controlled waters from Environment Agency and the application for discharge consent must comply with the procedures set out in the WRA 1991 Schedule 10 and the Control of Pollution (Applications, Appeals and Registers) Regulations 1996 (see Comments).
- Must pay the relevant fee for making an application for a new or revised consent to the Environment Agency.
- The applicant must publicise the application in a local newspaper and in the London Gazette and notify any relevant local authorities and water undertakers (see Disclosure).
- Environment Agency can attach conditions to discharge consent, which must be complied with. Typically, a discharge consent sets the legal limits on the type, concentration and total volume of the pollutants/material that may be discharged and includes various other conditions such as the place of discharge, the design/construction of the discharge outlet, sampling and monitoring requirements, record keeping and maintenance procedures.

- Agency can serve enforcement notice specifying measures to remedy any breach of consent.
- Where pollution occurs or is likely to occur, Agency can serve "works notice" on person who caused or knowingly permitted it, requiring them to carry out anti-pollution works and operations. Appeals against the issue of works notices can be made to Secretary of State within 21 days. The Anti-pollution Works Regulations 1999 (SI 1999 No.1006), prescribe the contents of anti-pollution works notices served under section 161A of the Water Resources Act 1991, the procedure to be followed in relation to appeals against such notices and the compensation for rights of entry in connection with anti-pollution works paid under section 161B of the Water Resources Act 1991.

DISCLOSURE

- The applicant must publicise the application in a local newspaper and in the London Gazette and notify any relevant local authorities and water undertakers. However, these publicity requirements may be dispensed with if the Agency considers that the discharge will have "no appreciable effect" on the receiving waters.
- Environment Agency is required to maintain registers of discharge consents and other information.
- There are certain restrictions on disclosure of information, including information deemed as commercially confidential or affecting national security.

COMPLEMENTARY LEGISLATION

- Surface Waters (River Ecosystem) (Classification) Regulations 1984.
- Environmental Protection Act 1990 Part I.
- Water Industry Act 1991.
- Urban Waste Water Treatment (England & Wales) Regulations 1994.
- Environment Act 1995.

- Surface Waters (Dangerous Substances) (Classification) Regulations 1997.
- Surface Waters (Dangerous Substances) (Classification) Regulations 1998 (see Comments).
- Pollution Prevention and Control Act 1999.
- Pollution Prevention and Control (England and Wales) Regulations 2000.
- Pollution Prevention and Control (Scotland) Regulations 2000.

RELATED EC LEGISLATION

- EC Directive 76/464 on pollution caused by Discharge of Certain Dangerous Substances into Aquatic Environment plus daughter Directives. *N.B.* There are numerous substance-specific "daughter" Directives.

COMMENTS

- On summary conviction, the penalty for failing to comply with S 85 can include a prison term of not more than three months or a fine not exceeding £20 000. Upon conviction on indictment, imprisonment for a term not exceeding two years may be imposed and/or an unlimited fine.
- Section 169 of the WRA 1991 provides powers of entry to an authorised person in connection with the investigation of pollution offences.
- The Surface Waters (Dangerous Substances) (Classification) Regulations 1998 (SI 1998 No. 389), which came into force on 25th March 1998, have set Environmental Quality Standards (EQSs) for a further 20 substances in England and Wales, following the 12 set by the Surface Waters (Dangerous Substances) (Classification) Regulations 1997 in autumn 1997. The Environmental Quality Standards (EQSs) will be taken into account by the environment agencies when issuing discharge consents under the Water Resources Act 1991.
- Storm water drainage is nearly always a problem area. Make sure that all the site drainage points are identified and that their routing to the receiving waters, and especially the owners of the route between the boundary and receiving water, are known.

RELATED DOCUMENTATION

- Application for consent and related correspondence.
- Notice of application.
- Discharge consent (under this Act or authorisation under EPA Part I or PPCR).
- Monitoring information.
- Any enforcement/variation/works notice served.

10.3 Discharges to Sewers

LEGISLATION

- Water Industry Act 1991.
- Water Industry Act 1999.

KEY DATES

- In force from 1/12/91.
- 1999 Act in force from 23/12/99 and 1/4/00 (see Comments).

PURPOSE

- Ensure supply of wholesome water and provision of sewage services by private sector water and sewerage undertakers.

APPROACH

- Duty on water undertaker to supply wholesome water and duty on sewerage undertaker to allow discharge of trade effluent to public sewer.
- Requires any site discharging trade effluent into sewers to have a consent from the appropriate water company. The consent is given after an application is submitted by the relevant facility and will be granted subject to certain conditions.

REGULATORY AUTHORITY

- Director General of Water Services.
- Environment Agency.
- Local authorities have supervisory functions in respect of provision of wholesome water supplies.

COVERAGE

- Trade effluent from industrial/trade premises discharged to sewer.
- Waterworks, mains, pipes.

COMPLIANCE REQUIREMENTS

- Prior consent or agreement to discharge trade effluent into public sewer required with sewerage undertaker and discharger must serve application notice on undertaker specifying various details about proposed discharge (see Comments).
- Sewerage undertaker may specify such conditions as nature, quantity, rate, monitoring of discharge.
- Most hazardous effluent prescribed as 'special category effluent' and subject to consent by Environment Agency.

DISCLOSURE

- Sewerage undertaker to maintain registers of trade effluent consents.
- Environment Agency can serve notice on any person requiring them to provide information for the purpose of its functions in relation to special category effluent.

COMPLEMENTARY LEGISLATION

- Trade Effluent (Prescribed Processes and Substances) Regulations 1989 (as amended).
- Environmental Protection Act 1990 Part I.
- Water Resources Act 1991.
- Urban Waste Water Treatment (England and Wales) Regulations 1994.
- Environment Act 1995.
- Water Supply (Water Quality) Regulations 1999 (as amended) (See Comments).
- Water Industry (Charges) (Vulnerable Groups) Regulations 1999.
- Water and Sewerage ((Conservation, Access and Recreation) (Code of Practice) Order 2000.
- The Drinking Water (Undertakings) (England and Wales) Regulations 2000.
- Pollution Prevention and Control (England and Wales) Regulations 2000.

RELATED EC LEGISLATION

- EC Directive 80/778 on Quality of Water for Human Consumption (Drinking Water Directive).
- EC Directive 91/271 on Urban Waste Water Treatment.

COMMENTS

- EPA Part I or PPCR enables EA to control discharges of trade effluent to sewers from prescribed processes or installations subject to IPC or IPPC by means of conditions in IPC authorisation or IPPC permit. However, such processes or installations will also require consent from the sewerage undertaker under the Water Industry Act 1991.
- Appeals against refusals of consent (or attached conditions) can be made to Director General of Water Services.
- Appeals in relation to special category effluent are dealt with by Environment Agency.
- Water Supply (Water Quality) (Amendment) Regulations 1999 (SI 1999 No. 1524), which came into force on 30th June 1999, amend the Water Supply (Water Quality) Regulations 1999 to ensure that drinking water supplies are not contaminated with *Cryptosporidium*, a parasite which is detrimental to human health. Water undertakers are required to carry out risk assessments in their treatment works to establish whether there is a "significant risk" from *Cryptosporidium* in water supplied for human consumption.
- Water Industry Act 1999 is concerned with the setting up of water charging schemes in England and Wales.
- Water Industry (Charges) (Vulnerable Groups) Regulations 1999 define vulnerable groups who are to be given assistance with water and sewerage charges by water and sewerage undertakers, and the nature of the assistance to be provided.
- The Water and Sewerage (Conservation, Access and Recreation) (Code of Practice) Order 2000 came into force on 27th March 2000 and gave legal authority to a Code of Practice. This Code gives practical guidance to water and sewerage undertakers and to the Environment Agency relating to their environmental and recreational duties under the Water Industry Act 1991 and the Environment Act 1995.

- The Drinking Water (Undertakings) (England and Wales) Regulations 2000 came into force on 14th June 2000 to comply with a judgement of a European Court in the case of *Commission v United Kingdom (1999)*. The Regulations specify requirements to be given where a contravention has occurred, and only if these are met can an undertaking by the water company be accepted as compliance with the Drinking Water Directive. Conditions include enforcing authority is satisfied that there is no potential danger to human health, remedial steps are most appropriate and there is no reasonable alternative methods of maintaining supply to the population.

RELATED DOCUMENTATION

- Application notice and related correspondence.
- Discharge consent.
- Monitoring information.
- Any information notice served.

10.4 Groundwater

LEGISLATION

- Groundwater Regulations 1998 (SI 1998 No. 2746).

KEY DATES

- In force from 1/1/99 (some) and 1/4/99 (remainder).

PURPOSE

- Complete the implementation of the Groundwater Directive (EC Directive 80/68) for England, Wales and Scotland.

APPROACH

- Supplement Regulation 15 of the Waste Management Licensing Regulations 1994 and existing water pollution legislation. The Regulations require the three regulatory authorities (see Regulatory Authority) to use both their new powers under the Regulations and their existing powers under Part II of the Control of Pollution Act 1974, Part III of the Water Resources Act 1991 and Part I of the EPA 1990:
  - To *prevent* the direct or indirect discharge of list I substances to groundwater.
  - To *control pollution* resulting from the direct or indirect discharge of list II substances.
- The new powers under the Regulations introduce:
  - A new requirement for an authorisation for the disposal, or tipping for the purposes of disposal, of list I or II substances in cases where a waste management licence under Part II of the Environmental Protection Act 1990 is not already required.
  - A new procedure for prohibiting or regulating by notice other activities in or on land which "might" lead to indirect discharge of List I substance to groundwater or to pollution by List II substance.

REGULATORY AUTHORITY

- Environment Agency.
- Scottish Environment Protection Agency (SEPA).
- The Secretary of State.

COVERAGE

- List I or II substances.
- There are exclusions from the Regulations in relation to radioactive substances, domestic effluent from isolated dwellings, small quantities and concentrations of list I or II substances, and activities for which a waste management licence is required (Regulation 2). There are also exceptions from the main controls (subject to certain safeguards) in relation to:
  - Discharges to groundwater unsuitable for other uses.
  - Water used for geothermal purposes or pumped out of mines, quarries or civil engineering works.
  - Artificial recharges of groundwater for management purposes.

COMPLIANCE REQUIREMENTS

- Must obtain authorisation for disposal or tipping of List I or II substances.
- Must pay application fee and annual subsistence charge for authorisations.
- Must comply with any notice served by Agencies in relation to other activities in or on land which "might" lead to indirect discharge of List I substance to groundwater or to pollution by List II substance. The notices may prohibit activity or authorise it subject to conditions.

DISCLOSURE

- Regulation 22 deals with information to be included on water pollution registers, which will include:
  - Copies of authorisations.

- Information provided to Agencies in connection with authorisations.
- Monitoring data.

COMPLEMENTARY LEGISLATION

- Control of Pollution Act 1974.
- Water Resources Act 1991.
- Environmental Protection Act 1990 Parts I and II.
- Waste Management Licensing Regulations 1994.
- Environment Act 1995.
- Pollution Prevention and Control (England and Wales) Regulations 2000.

RELATED EC LEGISLATION

- EC Directive 80/68 – Groundwater Directive.

COMMENTS

- Appeals can be made against Agencies' decisions on authorisations.
- In exercising their powers, Agencies will be adopting a priority-based procedure, based on vulnerability of groundwater concerned.

RELATED DOCUMENTATION

- Any relevant waste management licence.
- Copies of authorisations.
- Information provided to Agencies in connection with authorisations.
- Any notice served.
- Monitoring data.

10.5 Oil Storage

LEGISLATION

- Control of Pollution (Oil Storage) (England) Regulations 2001 (SI 2001 No. 2954)

KEY DATES

- In force from 1/3/02.

PURPOSE

- To prevent pollution of controlled waters by oil.

APPROACH

- Imposes certain standards on persons having custody or control of oil.

REGULATORY AUTHORITY

- Environment Agency.

COVERAGE

- England only.
- Applies to the storage of oil, including petrol, on any premises except:
  - Waste oil satisfying the definition in the Waste Management Licensing Regulations 1994.
  - When the oil is situated in a building or wholly underground.
  - Storage container having a capacity of 200 litres or less.
  - Premises used as a private dwelling and storage capacity 3500 litres or less.
  - Premises used for refining oil.
  - Premises that are a farm and the oil is used for agriculture under the Agriculture Act 1947.

COMPLIANCE REQUIREMENTS

- Oil shall be stored in a container which is of sufficient strength and structural integrity to ensure it is unlikely to burst or leak.
- The oil container must be situated within a secondary containment system which satisfies certain specified requirements.
- Additional specified requirements must be met for fixed tanks and mobile browsers.

DISCLOSURE

COMPLEMENTARY LEGISLATION

- Agriculture Act 1947.
- Water Resources Act 1991.
- Waste Management Licensing Regulations 1994.
- Building (Amendment) Regulations 2001.

RELATED EC LEGISLATION

COMMENTS

- Transitional provisions are specified and include that the Regulations do not apply until 1st September 2005 for oil stored in a container if that container were used for that purpose on any premises before 1st September 2001 and that the container is situated less than 10 metres from specified waters or less than 50 metres away from a well or borehole. If these distances are not satisfied then the Regulations apply to the container from 1st September 2003.
- The Environment Agency can, where the transitional provisions apply, issue a notice to minimise the risk of water pollution on the persons in control, or having custody of, the oil. They can only do this if they consider there is a serious risk of pollution of controlled waters. A person served with such a notice has the right of appeal to the Secretary of State.

- Breach is a criminal offence punishable by a fine.
- Building Regulations 2001 impose certain requirements relating to building new oil storage tanks and pipes.

RELATED DOCUMENTATION

- Site storage plans and any associated documentation.
- Any audit of storage facilities.
- Environment Agency notice and associated correspondence.

10.6 Prescribed Processes and Substances

LEGISLATION

- Trade Effluents (Prescribed Processes and Substances) Regulations 1989 (SI 1989 No. 1156).
- Trade Effluents (Prescribed Processes and Substances) (Amendment) Regulations 1990 (SI 1990 No. 1629).
- Trade Effluents (Prescribed Processes and Substances) Regulations 1992 (SI 1992 No. 339).

KEY DATES

- In force from 1/9/89.
- 1990 Regulations in force from 31/8/90.
- 1992 Regulations in force from 20/3/92.

PURPOSE

- Specify categories of effluent subject to control by Environment Agency when discharged to public sewers.

APPROACH

- Prescribe substances and processes as 'special category effluent' for which consent by Agency is required.
  Note: To avoid duplication of controls, trade effluent from any IPC or IPPC regulated process is not considered a special category effluent.

REGULATORY AUTHORITY

- Environment Agency.

COVERAGE

- Prescribed substances.
- Effluent derived from prescribed processes.

COMPLIANCE REQUIREMENTS

- General duty of sewerage undertakers to accept effluent does not apply to special category effluent. To allow discharge of special category effluent sewerage undertaker must refer application to Environment Agency.
- Environment Agency will determine whether consent prohibited or, if consent given, what conditions (such as nature, quantity, rate, monitoring of discharge) should be attached.

DISCLOSURE

- Sewerage undertaker to maintain registers of trade effluent consents.
- Environment Agency can serve notice on any person, requiring them to provide information for the purpose of its functions in relation to special category effluent.

COMPLEMENTARY LEGISLATION

- Environmental Protection Act 1990 Part I.
- Water Resources Act 1991.
- Urban Waste Water Treatment (England & Wales) Regulations 1994.
- Trade Effluent (Prescribed Processes and Substances) Regulations 1989 (as amended).
- Environment Act 1995.
- Pollution Prevention and Control (England and Wales) Regulations 2000.

RELATED EC LEGISLATION

- Framework EC Directive 76/464 on Pollution Caused by Discharge of certain Dangerous Substances into Aquatic Environment plus daughter directives.

COMMENTS

RELATED DOCUMENTATION

- Application notice and related correspondence.
- Discharge consent.
- Monitoring information.
- Any information notice served.

10.7 Urban Waste Water Treatment

LEGISLATION

- Urban Waste Water Treatment (England and Wales) Regulations 1994 (SI 1994 No. 2841).
- Urban Waste Water Treatment (Scotland) Regulations 1994 (SI 1994 No. 2842).

KEY DATES

- In force from 30/11/94.

PURPOSE

- Set down minimum standards for the treatment of urban waste water (*i.e.* domestic sewage and industrial waste waters).
- Modify UK legislation to comply with requirements of EC Directive 91/271 on Urban Waste Water Treatment.

APPROACH

- Places duty on sewerage undertakers to ensure that collecting systems for – and standard of treatment to – urban waste water complies with minimum standards set in Regulations by specified dates. These treatment standards, and the time-frame within which they must be met, vary according to the population of the urban area concerned. Secondary treatment is required by 31/12/00 for all towns with a population equivalent of more than 15 000. Stricter standards are required in sensitive areas and lower standards are required in less sensitive "high natural dispersion areas" (as classified by the Secretary of State).
- Places duty on sewerage undertakers to ensure that industrial waste water is pre-treated and trade effluent consents can be modified for this purpose.
- Imposes duty on Environment Agency to ensure that direct discharges of biodegradable industrial waste water from certain industrial sectors, *e.g.* milk, meat, brewing, alcohol, are subject to appropriate consent conditions after 31/12/00.

- Imposes ban on dumping of sewage sludge from ships from 31/12/98, as agreed at the Conference for the Protection of the North Sea in March 1990.

REGULATORY AUTHORITY

- Sewerage undertakers.
- Environment Agency.
- Director General of Water Services.
- Secretary of State.

COVERAGE

- Domestic sewage.
- Industrial waste water to sewer, *i.e.* trade effluent.
- Direct discharges of biodegradable industrial waste water from certain industry sectors.
- Sewage sludge.

COMPLIANCE REQUIREMENTS

- Ensure that industrial waste water discharged to sewer is pre-treated where so required in trade effluent consent (see also Comments).
- Comply with ban on dumping of sewage sludge from ships from 31/12/98 and the progressive reduction before that date of the total amount of toxic, persistent and bioaccumulable materials in sludge.

DISCLOSURE

- Environment Agency must keep maps and certificates relevant to Regulations, and make them available to public for inspection.

COMPLEMENTARY LEGISLATION

- Trade Effluent (Prescribed Processes and Substances) Regulations 1989 (as amended).

- Water Industry Act 1991.
- Water Resources Act 1991.

RELATED EC LEGISLATION

- EC Directive 91/271 on Urban Waste Water Treatment.

COMMENTS

RELATED DOCUMENTATION

- Discharge consents to sewer.
- Any correspondence with sewerage undertaker/Environment Agency relating to impact of Regulations.
- Maps and certificates relevant to classification of local area.

11 WASTE

11.1 Contaminated Land

LEGISLATION

- Environmental Protection Act 1990 Part IIA: Contaminated Land.
- Contaminated Land (England) Regulations 2000 (SI 2000 No. 227).
- Contaminated Land (Scotland) Regulations 2000 (SSI 2000 No. 178).
- Contaminated Land (Wales) Regulations 2001 (SI 2001 No. 2197).

KEY DATES

- In force from 1/4/00 (England only).
- In force from 14/7/00 (Scotland only).
- In force from 1/7/01 (Wales only).

PURPOSE

- Part IIA creates a new framework for the identification and remediation of contaminated land, which is underpinned by the new Regulations.

APPROACH

- Local authority must inspect area "from time to time" to identify contaminated land and land suitable for designation as "special sites" (which are subject to control by Environment Agency). The latter are now prescribed by the new 2000 Regulations.
- Establishes procedures for remediation, including consultation process and serving of remediation notice by relevant enforcing authority on "appropriate person(s)", specifying remediation

work to be carried out and the deadline for doing so. Procedures relating to the remediation notices are set out in the new 2000 Regulations.

REGULATORY AUTHORITY

- Local authority.
- Environment Agency (for special sites).

COVERAGE

- Contaminated land defined as land in such a condition, by reason of substances in, on or under land that (a) significant harm is being caused or significant possibility of such harm being caused; or (b) pollution of controlled waters, is being, or is likely to be caused.
- "Special sites" (more potentially polluting landfills) are prescribed by Regulations 2 and 3, and Schedule 1 of the new 2000 Regulations. These are sites upon which the refining of certain hydrocarbons, the manufacture of explosives and prescribed processes under IPC or local air pollution control have taken place, sites owned or operated by certain military functions, land adjacent to any of the above sites or land, or land contaminated by substances from such sites. Land which affects certain controlled waters is also a "special site", the meaning of "affect" being defined in section 78A(8) of the EPA, including where certain chemicals enter the water from such sites or the chemicals are contained in certain underground strata.

COMPLIANCE REQUIREMENTS

- Remediation notice served on "appropriate person" (person who caused or knowingly permitted substances to be present on land or alternatively the owner or occupier) will specify remediation steps to be carried out and deadline for compliance.
- Failure to comply with any requirement of remediation notice is an offence.

- Where remediation notice not complied with, regulatory authority entitled to carry out remediation and recover reasonable costs incurred.

DISCLOSURE

- Regulatory authority to maintain public registers containing remediation notices and statements, appeals against notices and charging notices, convictions for prescribed offences and various other particulars. Note that the new 2000 Regulations (Schedule 3) adds to the items which must be included on the register.
- Provisions exist for exclusion of information on grounds of national security or commercial confidentiality.

COMPLEMENTARY LEGISLATION

- EPA 1990 Part III.
- Water Resources Act 1991.
- Environment Act 1995.
- Planning and Compensation Act 1991 (Amendment of Schedule 18) (England) Order 2002 (SI 2002 No. 116).

RELATED EC LEGISLATION

COMMENTS

- Part IIA was inserted into the Environmental Protection Act 1990 by the Environment Act 1995, but was not brought into force until April 1st 2000. The new framework builds on the existing statutory nuisance legislation contained in Part III of EPA, and replaces it in respect of contaminated land.
- Right of appeal against remediation notices exists.
- SI 2002 No. 116 in force from 22nd February 2002 provides for compensation payable to a person who has to grant rights over their land to another party so that they can comply with a remediation order.

RELATED DOCUMENTATION

- Remediation notices and statements.
- Charging notices.
- Appeals against remediation and charging notices.
- Land surveys.
- Historical use of land in formation.

11.2 Control of Waste Carriers

LEGISLATION

- Control of Pollution (Amendment) Act 1989.
- Controlled Waste (Registration of Carriers and Seizure of Vehicles) Regulations 1991 (SI 1991 No. 1624).
- Controlled Waste (Registration of Carriers and Seizure of Vehicles) (Amendment) Regulations 1998 (SI 1998 No. 605).

KEY DATES

- In force from 16/7/91 (some) and 14/10/91 and 1/4/92 (remainder).
- In force from 14/10/91.
- (Amendment) Regulations 1998 in force from 1/4/98.

PURPOSE

- Prevent unregulated transport of waste.

APPROACH

- The 1989 Act requires the registration of waste carriers and makes it a criminal offence for anyone, other than the producer, who is not a registered carrier to transport controlled waste to or from anywhere in Great Britain. The Act also provides for the seizure of vehicles used for illegal waste disposal.
- The 1991 Regulations enact the requirements of the 1989 Act, specifying those carriers needing to register and the procedures for registration. In addition, they set out the procedure for seizure of vehicles used for illegal waste disposal.
- The (Amendment) Regulations 1998 amend the procedures for registration as a waste carrier, and provide for the fees payable to be those specified in a charging scheme made under the Environment Act 1995.

REGULATORY AUTHORITY

- Environment Agency.

COVERAGE

- Relates to the carriage of controlled waste (*i.e.* Directive waste) in the course of business or otherwise with a view to profit. Note: There are exemptions for certain types of carrier.
- Transport by air, land, sea.

COMPLIANCE REQUIREMENTS

- Carrier (including importer and exporter) required to register with Environment Agency to carry controlled waste on standard application form provided by the Agency. Non-registration is a criminal offence.
- Fees must be paid to Environment Agency. Fees payable are those specified in a charging scheme made under the Environment Act 1995.
- The registration must be renewed every 3 years.
- Must notify Environment Agency of any change in circumstances which would affect information entered in the register.
- Carrier under duty to produce authority to transport controlled waste if so required by Environment Agency officer/police.
- Environment Agency may revoke registration if convicted of prescribed offence/deemed undesirable to continue.
- Environment Agency may apply for a warrant to seize and remove vehicle.
- Note: As part of the duty of care introduced under Section 34 of the EPA 1990, a facility should check the carrier's registration by asking to see their Certificate of Registration or an official copy provided by the Environment Agency.

DISCLOSURE

- Register of carriers maintained by Environment Agency and is accessible to members of the public.

COMPLEMENTARY LEGISLATION

- Environmental Protection Act 1990 Part II.
- Waste Management Licensing Regulations 1994 (as amended).
- Transfrontier Shipment of Waste Regulations 1994.
- Environment Act 1995.

RELATED EC LEGISLATION

COMMENTS

- Waste producers should check that their waste is, in practice, disposed of in accordance with the Regulations, *e.g.* that it goes to the correct destination.

RELATED DOCUMENTATION

- Application.
- Registration.
- Any notification of change in circumstances.
- Any revocation.
- Carrier's registration.
- Certificate of registration.
- Waste details.

11.3 Duty of Care

LEGISLATION

- Environmental Protection (Duty of Care) Regulations 1991 (SI 1991 No. 2839).
- Environmental Protection Act 1990 Section 34.

KEY DATES

- In force from 1/4/92.

PURPOSE

- The 1991 Regulations serve to implement and supplement the duty of care introduced by Section 34 of the EPA 1990, which aims to ensure the safe management of controlled waste from generation to final disposal.

APPROACH

- Section 34 of the EPA introduced a new concept known as the "duty of care" as respects waste. This requires all those who are in any way engaged in handling waste (*i.e.* who import, produce, carry, keep, treat or dispose of controlled waste, or who as a broker has control of such waste) to take reasonable and appropriate steps in relation to it, otherwise they commit a criminal offence. This duty follows the waste from its point of production to its final disposal point, so that the producer's responsibility no longer stops at his/her gate.
- The 1991 Regulations impose additional responsibilities on those subject to the Duty of Care in relation to the making and keeping of records. When waste is transferred, the regulations require both the transferor and the transferee of waste to complete, sign and keep for at least two years a transfer note containing information about the waste and the parties to the transfer.

REGULATORY AUTHORITY

- Environment Agency.

COVERAGE

- Duty of care applies to all those who import, produce, carry, keep, treat or dispose of controlled waste, or who as a broker has control of such waste.
- The only exception relates to occupiers of domestic property regarding their domestic waste.
- Duty of care applies to controlled waste/Directive waste only.

COMPLIANCE REQUIREMENTS

- The S 34 Duty of Care imposes a number of specific requirements which the holder of waste should meet:
  - To take all the steps which are reasonable in the circumstances to ensure that the waste is disposed of safely and to prevent anyone else from carrying out any unauthorised deposit, treatment or disposal of the waste.
  - To prevent the waste escaping from their or anyone else's control.
  - To ensure that the waste is transferred only to an "authorised person" or to a person for "authorised transport purposes". Authorised persons include:
    - Waste collection authorities in England and Wales.
    - The holders of waste management licences or those exempted from having to obtain one.
    - Registered carriers of controlled waste or those exempted from having to register.
  - To ensure that a sufficient written description of the waste accompanies each transfer to enable person receiving the waste to avoid committing any of the offences and to comply with their duty to prevent escape.
- Under the 1991 Regulations, there must be a transfer note as well as a written description of waste, which must be checked and signed by both transferor and transferee. Transfer note must contain information specified by the 1991 Regulations, namely:

- A description of the waste, including quantity and type of container.
- The names, addresses, positions of both transferor (*i.e.* whether producer, importer, carrier *etc.*) and transferee.
- References to any necessary registrations including licence numbers.
- The name of the appropriate Agency.
- The date and place of transfer.

- Copies of transfer note and written description must be retained for period of 2 years from date of transfer by both transferor and transferee.

DISCLOSURE

- Copies of transfer note and written description must be retained for period of 2 years from date of transfer by both transferor and transferee, and this documentation must be presented to Environment Agency on request within this period.

COMPLEMENTARY LEGISLATION

- Environmental Protection Act 1990 Part II.

RELATED EC LEGISLATION

COMMENTS

- The Duty of Care is supported by a statutory *Code of Practice* which provides practical guidance for waste holders subject to the duty on how to discharge their duties. Breach of the Code of Practice is not an offence in itself, but a court may find that because someone did not comply with the Code of Practice, they failed to discharge their Duty of Care.
- Exemptions exist for repeated transfers of waste.

RELATED DOCUMENTATION

- Copies of transfer note and written descriptions, which must be retained for period of 2 years from date of transfer.

11.4 Landfill Tax

LEGISLATION

- Landfill Tax Regulations 1996 (SI 1996 No. 1527).
- Landfill Tax (Amendment) Regulations 1996 (SI 1996 No. 2100), (Amendment) Regulations 1998 (SI 1998 No. 61) 1999 (SI 1999 No. 3270) and 2002 (SI 2002 No. 1).
- Finance Act 1996.
- Landfill Tax (Qualifying Material) Order 1996 (SI 1996 No. 1528).
- Landfill Tax (Contaminated Land) Order 1996 (SI 1996 No. 1529).
- Landfill Tax (Site Restoration and Quarries) Order 1999 (SI 1999 No. 2075).
- Finance Act 2000.

KEY DATES

- Landfill tax levied from 1st October 1996.
- 1999 Order in force from 1/10/99.
- 1999 (Amendment) Regulations in force from 1/1/00.
- 2002 Regulations in force from 1/2/02.

PURPOSE

- Introduce a tax on the waste that goes to landfill in the UK so as to reflect the full environmental costs of disposing of waste to landfill.
- Its stated aim was firstly, to encourage business to produce less waste for disposal in landfills and to recover more value from the waste that is produced (*i.e.* through recycling); and secondly, to raise money for environmental remediation projects.

APPROACH

- The Finance Act 1996 (as amended), together with the Landfill Tax Regulations 1996 and other supporting regulations, introduces a landfill tax which is weight-based and levied per tonne of waste. The tax rates are £11 per tonne for standard waste (as from 1/4/00) and a reduced rate of £2 per tonne for specified inert wastes. The provisions, which came into force in October 1996, mean that the vast majority of waste disposed of in landfill sites is subject to a tax. The tax is paid by the operators of landfill sites and passed on indirectly to the waste producers.

- The impact of the tax is offset by cutting employers' National Insurance contributions so that it does not raise overall costs on business.
- Operators can benefit from tax rebate of up to 90% where they pay money into approved non-profit making environmental "bodies" which must spend money in furtherance of approved objects concerned with the environment, such as remediation, reclamation, recycling and promotion of the development of products from waste and markets for recycled waste. Such credits cannot exceed more than 20% of the operator's overall tax liability (See Comments).

REGULATORY AUTHORITY

- Customs and Excise.

COVERAGE

- Standard waste.
- Inert waste. The list of "qualifying materials" (*i.e.* inert or inactive wastes) attracting lower tax rate of £2 per tonne is contained in Landfill Tax (Qualifying Material) Order 1996. The materials are included in nine generic groups and include rocks/soils, ceramics/concrete, minerals, furnace slags, bottom ash/fly ash, low activity inorganic compounds, calcium sulphate, calcium hydroxide/brine and water containing other qualifying material in suspension.
- The Landfill Tax (Contaminated Land) Order 1996 provides that most forms of contaminated soil sent for disposal to landfill will be exempt from tax. Exemption will only apply where certificate is granted by Customs and Excise to persons carrying out a "qualifying" reclamation of the contaminated land in

question. The Landfill Tax (Site Restoration and Quarries) Order 1999, amends the Finance Act 1996 to provide two additional exemptions from landfill tax. The first exemption is for "qualifying material" which is used for the purposes of restoring a landfill site. The second exemption is for the disposal of "qualifying material" which is used to fill existing or former quarries. Again, certain conditions must be fulfilled.

- The 2002 Regulations provide that the temporary holding of material pending its use for site restoration purposes at the landfill site at which the disposal was made, or pending it being sorted, is not a taxable disposal. The material must be used for site restoration within three years from date of disposal or other period approved by Customs and Excise. A record of the material must be kept in the temporary disposal record for the disposal to be exempt.

COMPLIANCE REQUIREMENTS

- Landfill operators must notify Customs on form provided that they are carrying out taxable activities.
- Operators must submit quarterly tax returns on a further mandatory form, as well as maintaining a landfill tax account.
- Must make payments no later than the last day on which the returns must be submitted.
- Various records, including all invoices, transfer notes and the landfill tax account, must also be kept for a period of six years.

DISCLOSURE

COMPLEMENTARY LEGISLATION

RELATED EC LEGISLATION

COMMENTS

- Landfill Tax (Amendment) Regulations 1999 (SI 1999 No. 3270), which came into force on 1st January 2000, amend the main Landfill Tax Regulations 1996 to provide a number of changes to the scheme whereby landfill site operators are entitled to credit based upon the contributions they give to approved bodies with objects concerned with the environment.

RELATED DOCUMENTATION

- Notification to Customs on form provided that they are carrying out taxable activities.
- Quarterly tax returns on a further mandatory form.
- Various records, including all invoices, transfer notes and the landfill tax account which must be kept for a period of six years.

11.5 Litter

LEGISLATION

- Environmental Protection Act 1990 Part IV: Litter.

KEY DATES

- In force partly from 1/1/91: the rest at various commencement dates subsequently.

PURPOSE

- Introduce provisions for the control of litter by local authorities.

APPROACH

- Creates offence of leaving litter in a public open space.
- Places duty on local authority and certain other public bodies to keep land and highways clean and clear of litter.
- Regulatory authority may designate any land in their area as a litter control area, and take certain enforcement action, where the condition of the land is detrimental to the amenities of the locality.

REGULATORY AUTHORITY

- Local Authority.

COVERAGE

- Various public "relevant lands" and highways.
- Litter control areas.
- "Public open place" means a place in the open air to which the public are permitted to have access without payment.

COMPLIANCE REQUIREMENTS

- Must not throw down, drop or otherwise deposit litter in a public open space. Failure to comply is an offence. Local authority officer may issue a fixed penalty notice where such an offence committed.
- Regulatory authority may issue a litter abatement notice, where land within a litter control area is defaced by litter, which requires that the litter be cleared within a specified time or prohibiting the defacement by litter.

DISCLOSURE

COMPLEMENTARY LEGISLATION

- Litter Act 1983.
- Litter Control Areas Order 1991 (as amended).
- Litter (Fixed Penalty Notices) Order 1991.
- Litter (Fixed Penalty) (England) Order 2002.

RELATED EC LEGISLATION

COMMENTS

- 2002 Fixed Penalty Order in force from 1st April 2002 increased the fixed penalty from £25 to £50.

RELATED DOCUMENTATION

- Litter abatement notice.
- Fixed penalty notice.

11.6 Offshore Chemical Use or Discharge

LEGISLATION

- Offshore Chemicals Regulations 2002 (SI 2002 No. 1355).

KEY DATES

- In force from 15/5/02.

PURPOSE

- Implement the United Kingdom's obligations under the Convention for the Protection of the Marine Environment of the North-East Atlantic (OSPAR) Decision (2000/2) on a Harmonised Mandatory Control System for the Use and Reduction of the Discharge of Offshore Chemicals in relation to offshore activities.
- To protect the marine environment in the United Kingdom's territorial sea.

APPROACH

- Operators must obtain a permit to use and discharge chemicals.

REGULATORY AUTHORITY

- Secretary of State.

COVERAGE

- "Use" of chemicals defined as any application in connection with offshore activities that might result in a discharge.
- Offshore activities means any activities covered under the Petroleum Act 1998.
- Offshore chemicals means any chemical intentionally used in connection with offshore activities.

- Area covered is:
  - Those parts of the sea adjacent to England and Wales from the lower water mark to the landward baseline of the United Kingdom territorial sea.
  - The United Kingdom territorial sea apart from those areas comprised in Scottish controlled waters.
  - Specified areas anywhere designated under Section 1(7) of the Continental Shelf Act 1964.

COMPLIANCE REQUIREMENTS

- Must not use or discharge any offshore chemical in the relevant area otherwise than in accordance with the terms of, and conditions attached to a permit.
- Must publicise the permit application. This includes making the permit application available for public inspection and publishing in certain newspapers details of the application and where it can be inspected.
- Must inform Secretary of State with details of any accident or incident involving an offshore chemical where there has been:
  - Breach of the permit or its conditions.
  - Or may be, any significant effect on the environment.
- Must pay fees.

DISCLOSURE

- Publicity related to application.
- Permit Register including any monitoring results.

COMPLEMENTARY LEGISLATION

- Continental Shelf Act 1964.
- Petroleum Act 1998.
- Pollution Prevention and Control Act 1999.
- The Offshore Installations (Emergency Pollution Control) Regulations 2002.

RELATED EC LEGISLATION

COMMENTS

- These Regulations are made under Section 2 of the Pollution Prevention and Control Act 1999.
- The Regulations deal with the procedure for granting permits, the conditions of permits and the requirements for permit applications.
- The permit may be time limited or not. Provisions are stated for renewal of the permit, variation, revocation and surrender of permits.
- The Secretary of State must keep a register of permits.
- Inspectors can be appointed who have wide powers and duties which are specified.
- There is a right of appeal against the Secretary of State's decision.
- Individual liability is provided for in certain circumstances.
- The Offshore Installations Regulations 2002 aim to prevent and reduce pollution, and the risk of pollution following an accident involving an offshore installation.

RELATED DOCUMENTATION

- Permit application and related correspondence.
- Permit and any terms and conditions.
- Monitoring results.
- Publicity information.
- Breach or significant environmental effect information provided to Secretary of State.

11.7 Packaging Waste

LEGISLATION

- Producer Responsibility Obligations (Packaging Waste) Regulations 1997 (SI 1997 No. 648).
- Producer Responsibility Obligations (Packaging Waste) (Amendment) Regulations 1999 (SI 1999 No. 1361), (Amendment) (No. 2) Regulations 1999 (SI 1999 No. 3447).
- Producer Responsibility Obligations (Packaging Waste) (Amendment) (England and Wales) Regulations 2000 (SI 2000 No. 3375).
- Producer Responsibility Obligations (Packaging Waste) (Amendment) (England) Regulations 2002 (SI 2002 No. 732).

KEY DATES

- In force from 6/3/97.
- 2000 Amendment Regulations in force from 31/12/00.
- 2002 Amendment Regulations in force from 21/3/02.

PURPOSE

- Place a "producer responsibility" obligation on certain businesses to recover and recycle certain percentages of packaging waste.
- Meet the minimum targets imposed by EC Directive 94/62 on Packaging and Packaging Waste, namely 50% recovery and 15% recycling for each material by 2001.

APPROACH

- Imposes "shared responsibility" obligation on four sectors of packaging chain (1) packaging raw material manufacturers (2) converters (3) packers/fillers (4) retailers, to recover and recycle specific tonnages of packaging waste.
- Impose gradually increasing recovery and recycling targets, to meet, and then surpass, the minimum targets imposed by EC Directive 94/62. 2000 Amendment Regulations have now increased the targets for 2001 to 56% for recovery and 18% for recycling and for any subsequent years 59% and 19% respectively.

- Obligated businesses can either carry out the recovery and recycling obligation themselves, or join a registered 'exemption' compliance scheme. Such schemes will assume responsibility for meeting recovery and recycling targets on behalf of their members, who will be exempt from their legal obligation. The scheme operates in practice by compelling obligated businesses to provide evidence of compliance from reprocessors who recover and recycle packaging materials in the form of packaging recovery notes (PRNs).

REGULATORY AUTHORITY

- Environment Agency.

COVERAGE

The Regulations apply to all businesses:

- Which act as either (1) packaging raw material manufacturers (2) converters of raw materials into packaging (3) packers or fillers of packaging (4) retailers or (5) importers (but not exporters or wholesalers) of packaging waste.
- Which own the packaging involved.
- Which handle more than 50 tonnes of packaging waste (materials) a year and which supply on to another stage in the packaging chain or to the final user.
- With a turnover of over £2 million.

COMPLIANCE REQUIREMENTS

- Must register with Agency, and pay the appropriate fee by 7th April each year.
- Must provide data on obligated packaging handled in previous year to Agency by 7th April each year.
- Must calculate legal recycling and recovery obligations and meet them. The recycling and recovery obligations for a business are calculated as follows:
  - Multiply the tonnage of recyclable material (packaging), handled by the business, by the percentage obligation

associated with each activity, (packaging raw material manufacturers (6%), converters (9%), packers/fillers (37%), retailers (48%), by the recycling target for the relevant year.

- Producers with over £2 million turn over and applying for individual registration with Environment Agency must also now include an operational plan showing the steps to be taken to comply with their obligations.
- Alternatively, if joining a compliance scheme, must check that it is registered. Must also meet the conditions of membership, such as payment of fee and provision of data.
- All obligated businesses must demonstrate compliance in any year by sending a certificate of compliance to the Environment Agency which states that the organisation met its obligation the previous year.

DISCLOSURE

COMPLEMENTARY LEGISLATION

- Environment Act 1995, Section 93, under which the Regulations are introduced.
- The Packaging (Essential Requirements) Regulations 1998.

RELATED EC LEGISLATION

- EC Directive 94/62 on Packaging and Packaging Waste.

COMMENTS

- Note: Exports are excluded from the scope of the Regulations.
- Note: The DETR has published an accompanying document, '*Forward Look for Planning Purposes*' to help businesses take a strategic approach to meeting the new recovery and recycling targets.
- It is a criminal offence to fail to take reasonable steps to meet the recovery/recycling obligations.

RELATED DOCUMENTATION

- Registration with Agency (compliance scheme).
- Data on obligated packaging handled in previous year.
- Operational plan (if appropriate).
- Data calculating legal recycling and recovery obligations.
- Certificate of compliance (PRNs).

11.8 Packaging – Essential Requirements

LEGISLATION

- The Packaging (Essential Requirements) Regulations 1998 (SI 1998 No. 1165).

KEY DATES

- In force from 31/5/98.

PURPOSE

- Further implement the 1994 EC Directive 94/62 on packaging waste by transposing its provisions relating to the "essential requirements" to be satisfied by packaging as well as limits on its heavy metal content.

APPROACH

- Imposes a duty on a "responsible person" (essentially packer/fillers or importers) who places packaging on the market to ensure that it satisfies the relevant "essential requirements" set out in the Schedule.
- Sets out a phased programme for the reduction of heavy metals present in packaging, and imposes a duty on "responsible person" to comply with these limits.

REGULATORY AUTHORITY

- Weights and measures authorities in Great Britain.

COVERAGE

- Heavy metal concentration limits do not apply to packaging which is made entirely of lead crystal glass.

COMPLIANCE REQUIREMENTS

- "Responsible person" who places packaging on the market is under a duty to ensure that it satisfies the relevant "essential requirements" set out in the Schedule (Regulation 6). These requirements – set out in fairly general terms – stipulate that packaging must be reusable or recoverable and that the volume and weight of packaging must be limited to the "minimum adequate amount" to maintain the necessary level of safety, hygiene and acceptance for both the packed product and consumer. Packaging will satisfy these requirements if it complies with national standards which implement harmonised European standards (being drafted by the European standards body CEN).
- "Responsible person" who places packaging on the market is under a duty to ensure that the sum of the concentration levels of lead, cadmium, mercury and hexavalent chromium of packaging or any of its components does not exceed (Regulations 6,7,8):
  - 600 parts per million by weight after 30th June 1998,
  - 250 ppm after 30th June 1999 and
  - 100 ppm after 30th June 2001.
- "Responsible person" must keep technical documentation on compliance with both the essential requirements and heavy metals concentration limits, and to make this available at the request of the enforcement authority for a period of four years from the date the packaging is placed on the market (Regulations 11 and 12).

DISCLOSURE

- Technical documentation given to enforcement authority.

COMPLEMENTARY LEGISLATION

- Producer Responsibility Obligations (Packaging Waste) Regulations 1997.

RELATED EC LEGISLATION

- EC Directive 94/62 on Packaging and Packaging Waste.

COMMENTS

- Regulation 15 provides for penalties for breaches of the Regulations. A defence of due diligence is also available (Regulation 16) while proceedings can also be taken against persons other than the "principal offender" where their act or default contributed to a breach of the Regulations (Regulation 17).

RELATED DOCUMENTATION

- Technical documentation on compliance with both the essential requirements and heavy metals concentration limits.

11.9 Special Waste

LEGISLATION

- Special Waste Regulations 1996 (SI 1996 No. 972).
- Special Waste (Amendment) Regulations 1996 (SI 1996 No. 2019) and (Amendment) Regulations 1997 (SI 1997 No. 251), 2001 (SI 2001 No. 3148), 2001 (SI 2001 No. 3545 – Wales only).

KEY DATES

- In force from 1/9/96.
- 3148 Regulations in force from 1/11/01 except Regulation 10 (changes to prescribed form of consignment note) from 1/12/01 and Regulation 11 (new prescribed form of carrier's schedule) from 1/5/02. 3545 Regulations same dates as for 3148 but apply to Wales only.

PURPOSE

- Regulate the management and handling of particularly dangerous/hazardous waste or waste which is difficult to dispose of.
- These 1996 Regulations, which repeal and replace the Control of Pollution (Special Waste) Regulations 1980, serve to implement the EC Directive 91/689 on Hazardous Waste.

APPROACH

- The 1996 Regulations provide a new definition of special waste, which implements the definition of hazardous waste contained in the EC Directive 91/689 on Hazardous Waste. To implement the Directive, the Regulations also adopt the criteria for determining special waste contained in the EC Hazardous Waste List (Council Decision 94/904/EC), which has led to an increase in the number of waste types now classified as a special waste.
- Set out pre-notification and consignment procedure for management and handling of special waste. Under the Regulations, the movement and disposal of special waste

requires pre-notification of the proposed movement to the Environment Agency and use of a consignment note system which ensures that transfers of waste are properly described and documented. A unique code for the consignment must also be obtained from the Agency, and the code must be shown, together with the other information required, on the consignment notes which must accompany the waste when transported.
- The mixing of special waste with other waste or other categories of special waste is prohibited except where authorised under certain other waste management legislation.

REGULATORY AUTHORITY

- Environment Agency.

COVERAGE

- Special waste, *i.e.* controlled waste which falls within the definition of hazardous waste in the EC Directive 91/689 on Hazardous Waste and in particular the EC Hazardous Waste List (Council Decision 94/904).
- The definition also extends to certain other waste considered by the UK to display certain hazardous properties, including medicinal products only available by prescription.
- Certain radioactive wastes are also included.
- Household waste is excluded from the definition of special waste.
- Note: The Regulations do not apply to transfrontier shipments of waste covered by EC Regulation 259/93/EC.

COMPLIANCE REQUIREMENTS

- Producers/consignors of special waste must complete five copies of consignment note, which must contain unique code obtained from Environment Agency.
- Producers/consignors must pre-notify Agency, sending one copy of consignment note to the local Agency office for the area to

which the waste is being consigned at least 3 days before special waste is removed from premises (see Comments).
- Must pay fee to the Environment Agency for each consignment.
- Must comply with prohibition on mixing of special waste with other waste or other categories of special waste except where authorised under waste management licence (or IPC authorisation or IPPC permit).
- Carriers and disposers/consignees must also complete parts of the consignment note which accompanies the waste when transported.
- Those consigning and carrying special waste are both required to keep various records, including the consignment notes, for at least 3 years.
- Those disposing of special waste on land must record the location of each deposit and keep such records until their waste management licences are surrendered or revoked, then it must be handed to Agency.

DISCLOSURE

COMPLEMENTARY LEGISLATION

- EPA 1990 Part II.
- Waste Management Licensing Regulations 1994 (as amended).
- Transfrontier Shipment of Waste Regulations 1994.

RELATED EC LEGISLATION

- EC Directive 91/689 on Hazardous Waste.
- EC Hazardous Waste List (Council Decision 94/904).

COMMENTS

- Detailed guidance on the Regulations is contained in DETR Circular (6/96).
- Consignment note for special waste obviates the need for a duty of care transfer note as required by s 34 of the EPA 1990.

- Failure to comply with the Regulations is a criminal offence.
- Amending Regulations 1996 make minor changes to the definition of special waste and also various minor changes to the administration of the new regime.
- 2001 Amendment Regulations make minor amendments including that for second and subsequent consignments, in a succession of special waste consignments, code identifying first consignment must be entered on consignment note as well as relevant code for consignment in question and where a consignment is rejected any carrier's schedule forwarded to agency indicates which loads are not accepted. Regulation 10 amends the prescribed form of consignment notes and Regulation 11 provides a new carrier's schedule including additional information (postcode of the address of origin and the time of the carrier's signature) and removal of a description of waste.

RELATED DOCUMENTATION

- Registers containing consignment notes, which must be kept by consignors/producers, carriers and anyone who deposits special waste for at least 3 years.
- Disposers of special waste must also site records of location of deposit of special waste on land until waste management licence surrendered/revoked, then it must be handed to Agency.

11.10 Transfrontier Shipments

LEGISLATION

- EC Regulation 259/93 on the Supervision and Control of Shipments of Waste within, into and out of the Community.
- Transfrontier Shipment of Waste Regulations 1994 (SI 1994 No. 1137).

KEY DATES

- In force 6/5/94.

PURPOSE

- EC Regulation, which is directly applicable in the UK, seeks to regulate international shipments of waste.
- UK Regulations implement and supplement certain provisions of EC Regulation concerned with administration, insurance and enforcement.

APPROACH

- Where shipments of waste for disposal, notification must be made to competent authorities of exporting, importing and transit countries, and consent must be received.
- Where shipments of waste for recovery, wastes are classified in red, amber and green lists – according to their hazard rating – and control procedures vary accordingly.

REGULATORY AUTHORITY

- Environment Agency is competent authority for import and export of waste.
- Secretary of State is competent authority for transit.

COVERAGE

- All wastes.

COMPLIANCE REQUIREMENTS

- Where shipments of waste for disposal, notification must be made to competent authorities of exporting, importing and transit countries. Notifier must receive written consent from importing state before shipment takes place and enter into contract with disposer specifying environmentally sound disposal of waste.
- Where wastes destined for recovery, shipments of Red List wastes require prior written consent, Amber wastes are subject to simplified control procedure and Green wastes require no notification/consent.
- Importers/exporters of waste must apply to Agency for certificate demonstrating they have financial guarantee or other insurance sufficient to meet the cost of the shipment.

DISCLOSURE

- Agency required to establish registers of dealers and brokers.

COMPLEMENTARY LEGISLATION

- Transfrontier Shipment of Radioactive Waste Regulations 1993.

RELATED EC LEGISLATION

- EC Regulation on Supervision and Control of Shipments of Waste within, into and out of the Community (259/93) (*N.B.* this is directly applicable law in the UK).

COMMENTS

- Regulations repeal and replace the Transfrontier Shipment of Hazardous Waste Regulations 1988 (as amended).

RELATED DOCUMENTATION

- Certificate from Agency demonstrating financial guarantee or other insurance sufficient to meet the cost of the shipment.
- Where shipments of waste are for *disposal*:
  - Notifications to competent authorities of exporting, importing and transit countries.
  - Written consent from importing state.
  - Contract with disposer specifying environmentally sound disposal of waste.
- Where shipments of waste are for *recovery*:
  - Prior consent/notification according to whether waste is on red/ amber/green list.

11.11 Waste Management

LEGISLATION

- Environmental Protection Act 1990 Part II: Waste on Land.
- Controlled Waste Regulations 1992 (SI 1992 No. 588).
- Controlled Waste (Amendment) Regulations 1993 (SI 1993 No. 556).
- Environment Act 1995.

KEY DATES

- Provisions relating to regulatory and operational functions of local authorities in force from 1991.
- Section 34 Duty of Care in force from 1/4/92.
- Waste management licensing provisions in force from 1/5/94.

PURPOSE

- Introduce new framework for regulating the handling and disposal of waste from 'cradle to grave' so as to prevent/minimise pollution of environment and harm to human health.

APPROACH

- Sets out new criminal offence of treating, keeping or disposing of controlled waste in a manner likely to cause pollution of environment or harm to human health. This operates independently of licensing system.
- Sets out new system of waste management licensing to replace waste disposal licensing provisions of Control of Pollution Act 1974. Under the EPA 1990, a waste management licence is required for all sites which keep, treat, or dispose of controlled waste in or on specified land, or treat or dispose of waste by means of mobile plant.
- Imposes new "duty of care" on anyone who imports, produces, carries, keeps, treats, or disposes of controlled waste or brokers with control of such waste, to take reasonable and appropriate steps in relation to it, otherwise they commit a criminal offence.

- Separated regulatory and operational/disposal functions of the former Waste Disposal Authorities (WDAs) by creation of waste regulation authorities (whose powers have now been transferred to the Environment Agency) and arms-length waste disposal companies, subject to competitive tendering from private sector.
- Sets out provisions designed to encourage recycling of waste, and also powers enabling stricter controls for special waste.

REGULATORY AUTHORITY

- Environment Agency.
- Waste Collection Authority (WCA).

COVERAGE

- The definition of "Waste" in the EPA 1990 was amended by the Environment Act 1995 to refer to the definition of "Directive waste" laid down in the amended EU Framework Directive on Waste (91/156/EEC). Two questions must now be asked in determining whether a substance or object is waste or Directive waste:
  - Is it contained in the categories set out in Schedule 2B (which reproduces Annex I of the Waste Directive 75/442 as amended); *and*
  - Has the producer/holder discarded it or does he intend or is he required to discard it?
- However, the EPA primarily refers to the disposal of "controlled waste", including household, industrial and commercial waste. The Controlled Waste Regulations 1992 define the types of waste to be treated as household, industrial or commercial waste. However, the 1994 Waste Management Regulations amended the definition of controlled waste to provide that waste which is not Directive waste shall not be treated as controlled waste. As a result, the terms "controlled waste" and "Directive waste" are in effect interchangeable.

COMPLIANCE REQUIREMENTS

- Must not contravene criminal offence of treating, keeping or disposing of controlled waste in a manner likely to cause pollution of environment or harm to human health.
- Must obtain waste management licence to treat, keep or dispose of controlled waste on land (see *Waste Management Licensing: Waste Management Licensing Regulations 1994* (as amended) for greater detail)
  - Must comply with conditions specified by Agency in licence.
  - Must pay licence fee to Environment Agency.
  - Applicant must be "fit and proper person".
  - Must obtain planning permission for land in question.
  - Must apply to Agency for surrender or transfer of waste management licence.
  - Agency may revoke or suspend a waste management licence where an offence or continuation would lead to pollution/ harm to human health.
- Must comply with duty of care *i.e.* any person who imports, produces, carries, keeps, treats or disposes of controlled waste or brokers with control of such waste must take all reasonable measures to ensure its safe and authorised management and disposal (see *Duty of Care: Environmental Protection (Duty of Care) Regulations 1991* for greater detail).

DISCLOSURE

- Public register to be maintained by Agency with exceptions relating to information deemed as commercially confidential at the authority's determination.

COMPLEMENTARY LEGISLATION

- Environmental Protection Act 1990 Part II replaces most of controls over waste contained in the Control of Pollution Act 1974.
- Environmental Protection Act 1990 Part IV, relating to the control of litter.

- Environmental Protection (Duty of Care) Regulations 1991.
- Waste Management Licensing Regulations 1994 (as amended).
- Environment Act 1995, which significantly amends EPA Part II.
- Pollution Prevention and Control (England and Wales) Regulations 2000.

RELATED EC LEGISLATION

- Framework EC Directive on Waste 75/442 amended by 91/156 plus daughter directives.

COMMENTS

- Waste disposal contractors may be required to recycle waste and accept specified waste.
- Note: Agency can require removal of unlicensed deposits of controlled waste and, in certain cases, take remedial action charging expenses to the owner.
- Some waste disposal facilities require an IPPC permit as they come within the specified activities for which an installation requires a permit. The activities specified are disposal of waste by landfill, incineration or other means. However they have to dispose of a minimum specified quantity per day to require an IPPC permit.

RELATED DOCUMENTATION

- Applications and related correspondence.
- Licences.
- Planning permission.
- Relevant Waste Management Industry Training and Advisory Board (WAMITAB).
- Any variation/revocation/suspension notices served.
- Registration of any exemption.
- Applications for surrendering/transferring licences.
- Documentation relating to duty of care *i.e.* transfer notes and written descriptions of waste.

11.12 Waste Management Licensing

LEGISLATION

- Waste Management Licensing Regulations 1994 (SI 1994 No. 1056).
- Waste Management Licensing (Amendment) Regulations 1995 (SI 1995 No. 288) and (Amendment No. 2) Regulations 1995 (SI 1995 No. 1950), (Amendment) Regulations 1996 (SI 1996 No. 1279), (Amendment) Regulations 1997 (SI 1997 No. 2203), (Amendment) Regulations 1998 (SI 1998 No. 606) Amendment Regulations 2002 (SI 2002 No. 674) (see Comments).
- Environment Act 1995.
- Waste Management Regulations 1996 (SI 1996 No. 634).
- Waste Management (Miscellaneous Provisions) Regulations 1997 (SI 1997 No. 351).

KEY DATES

- In force from 1/5/94.
- (Amendment) Regulations 1998 in force from 1/4/98.
- (Amendment) Regulations 2002 in force from 14/3/02.

PURPOSE

- Implement waste management licensing system for waste handling, transfer and disposal sites contained in Part II of EPA 1990.
- Modify UK waste legislation to comply with requirements of EC Framework Directive on Waste (75/442 as amended).

APPROACH

- Under the EPA 1990 Part II, a waste management licence is required for all sites which "treat, keep or dispose" of controlled waste on land, or by means of mobile plant. In defining the applicability of this requirement, and implementing the amended Framework Directive, the 1994 Regulations limit the phrase "treatment, keeping and disposal" of waste to Directive disposal and recovery operations (see Coverage).

- The 1994 Regulations also implement the new waste management licensing system by:
  - Outlining the form and content of applications in respect of licences.
  - Specifying cases where a waste management licence is not required.
  - Defining the requirement that licensees be 'fit and proper persons' (set out in s 74 of the EPA 1990), and in particular the requirement for licensed waste facilities to be in the hands of "technically competent" persons.
  - Setting out appeals procedure and information to be include on registers.

REGULATORY AUTHORITY

- Environment Agency.
- Various statutory consultees (including HSE) must be consulted by Agency if it proposes to issue a licence.
- Waste Management Industry Training and Advisory Board (WAMITAB).

COVERAGE

- Licensing regime applies to controlled waste/Directive waste only.
- Directive disposal operations (set out in Part III of Schedule 4) includes treatment of waste prior to disposal, incineration on land and at sea, landfill, and permanent storage of waste, *e.g.* in a mine.
- Directive recovery operations (set out in Part IV of Schedule 4) includes reclamation or recycling of a variety of substances and the use of substances obtained from recycling or reclamation.
- Exemptions from waste management licensing (set out in Schedule 3) range from various recovery/recycling activities to burying of waste as fuel, use of certain wastes on land and activities that are regulated under other control regimes, *i.e.* IPC processes under Part I of EPA 1990 (except final deposit of waste) or IPPC permits under PPCR.

COMPLIANCE REQUIREMENTS

- Applications for waste management licences must include the appropriate information in the manner prescribed on standard application form provided by the Environment Agency. Application must also be accompanied by such other information as the Agency "reasonably requires", otherwise it can be refused.
- Must pay licence fee to Environment Agency specified in a charging scheme made under the Environment Act 1995.
- Applicant should be "fit and proper person", otherwise their licence may be refused:
  - Applicants or other relevant person must not be convicted of "relevant offence" (see Comments).
  - Manager of site must hold prescribed technical qualifications (WAMITAB certificates) – subject to transitional exceptions – in order to be considered "technically competent" (see Comments).
  - Applicant must have financial provision/insurance adequate to discharge obligations under the licence.
- Must obtain planning permission for land in question, otherwise licence will be refused.
- Must comply with conditions attached to licence.
- Must apply to Agency for surrender or transfer of waste management licence.
- Agency may revoke or suspend a waste management licence where an offence or continuation would lead to pollution/harm to human health.
- Activities exempt from waste management licensing must register with Environment Agency and pay fees specified in a charging scheme made under the Environment Act 1995.
- Applications for surrendering or transferring licences must include the appropriate information in the manner prescribed.
- Note: As part of the duty of care provisions introduced under the EPA, waste producers should ensure that waste is disposed of at a properly licensed site.

DISCLOSURE

- Register maintained by Agency must include information specified by the Regulations, including all applications, licences granted, monitoring information and variation/revocation/suspension notices.

COMPLEMENTARY LEGISLATION

- Control of Pollution Act 1974.
- Environmental Protection Act 1990.
- Controlled Waste (Registration of Carriers and Seizures of Vehicles) Regulations 1991 (as amended).
- Controlled Waste Regulations 1992 (as amended).
- Environment Act 1995.
- Pollution Prevention and Control (England and Wales) Regulations 2000.
- Pollution Prevention and Control (Scotland) Regulations 2000.
- Landfill (England and Wales) Regulations 2002.
- Offences under numerous other acts and regulations are 'relevant offences' for purposes of fit and proper persons requirement.

RELATED EC LEGISLATION

- Framework EC Directive on Waste 75/442 (as amended).

COMMENTS

- Reference should be made to Waste Management Paper 4 *Licensing of Waste Facilities*.
- "Relevant offences" are set out in Regulation 3 of 1994 Regulations. The (Miscellaneous Provisions) Regulations 1997 prescribe as additional "relevant offences" various criminal offences in relation to landfill tax.
- Amendment Regulations 1997 amend and replace Table 1 of the 1994 Regulations which specifies the type of certificates awarded by WAMITAB that individuals must obtain to show that they are technically competent for particular types of facility.

- Amendment Regulations 1998 amend procedures for registration with Environment Agency of activities exempt from waste management licensing and also waste brokers.
- Amendment Regulations 2002 enables the Environment Agency to authorise mobile plant for the recovery of ozone depleting substances from waste refrigeration equipment, fire extinguishers and other items.
- Landfill Regulations 2002 brings all landfills covered by these Regulations into the IPPC regime and this has resulted in some landfills transferring from a Waste Management Licence to an IPPC permit.

RELATED DOCUMENTATION

- Applications and related correspondence.
- Licences.
- Planning permission.
- Relevant WAMITAB certificates.
- Any variation/revocation/suspension notices served.
- Registration of any exemption.
- Applications for surrendering/transferring licences.

12 EXTERNAL NOISE

12.1 Miscellaneous

LEGISLATION

- Control of Pollution Act 1974 Part III: Noise.
- Control of Noise (Measurement and Register) Regulations 1976 (SI 1976 No. 37).
- Control of Noise (Codes of Practice for Construction and Open Sites) (England) Order 2002 (SI 2002 No. 461).

KEY DATES

- In effect from 1/1/76 (some) and then incrementally from that date.
- 2002 Order in force for England only 25/3/02.

PURPOSE

- Control noise on or from building, construction sites and other miscellaneous sources through both preventative and remedial provisions.

APPROACH

- Establish powers for local authorities to control noise and vibration on or from demolition and construction sites, including by means of serving a notice imposing requirements on person responsible for the construction works to observe various specified controls to minimise noise.
- Provide local authorities with powers to designate Noise Abatement Zones by noise abatement order, specifying classes of premises in which noise can be more strictly regulated. Local authority must measure noise emissions from all the premises in the zone within the specified classes and place information in public register.

- Establish powers for Secretary of State to issue Codes of Practice giving guidance on Part III of COPA and create regulations controlling noise emissions from plant and machinery.

REGULATORY AUTHORITY

- Local authority.

COVERAGE

- Sites, structures or roads where construction, demolition or repairs taking place.
- Noise Abatement Zones designated by local authority.
- Miscellaneous noise sources including loudspeakers in streets.

COMPLIANCE REQUIREMENTS

- Local authority can serve notice imposing requirements on person responsible for the construction works to observe various specified controls to minimise noise, having regard to the concept of best practicable means (see Comments). Non-compliance is an offence.
- Operator can apply for prior consent for work on construction sites which will serve as defence to above action by the local authority. Sites should apply for consent for operations prior to commencement of works.
- Exceeding the noise levels recorded in the register for Noise Abatement Zones without written consent from the local authority is an offence. Local authority can attach various conditions to consent.
- Comply with four Codes of Practice on basic information and procedures for noise and vibration control, surface coal extraction by opencast methods, noise and vibration control applicable to piling operations and surface mineral extraction, except coal, sites.

DISCLOSURE

- For Noise Abatement Zones the local authority must set up public noise register containing information on noise emissions from all the premises within the zone that are of a class specified in the noise abatement order.

COMPLEMENTARY LEGISLATION

- Control of Noise (Measurement and Register) Regulations 1976.
- Environmental Protection Act 1990 Part III.
- Noise and Statutory Nuisance Act 1993.

RELATED EC LEGISLATION

COMMENTS

- In issuing notice to construction sites, local authorities must also have regard to BS 5228 – Noise Control on Construction and Open Sites.
- Control of Noise (Measurement and Register) Regulations 1976 set out methods of measuring and recording noise levels for the purpose of the public noise register.
- For general guidance see DoE Circular 2/76: Implementation of Part III of Control of Pollution Act 1974.
- The four Codes of Practice are BS 5228 Part 1 1997, Part 3 1997, Part 4 1992 and Part 5 1997 available from the British Standards Institution.

RELATED DOCUMENTATION

- Prior consent for work on construction sites.
- Any notice served by local authority.
- Any written consent to exceed the noise levels recorded in the register for Noise Abatement Zones.

12.2 Noise in the Street

LEGISLATION

- Noise and Statutory Nuisance Act 1993.

KEY DATES

- In force from 5/1/94.

PURPOSE

- Amend and extend existing noise legislation (notably the statutory nuisance provisions in Part III of the EPA 1990) so as to address the problems of noise in the street from vehicles, machinery and audible intruder alarms.

APPROACH

- Adds noise in the street from vehicles, machinery or equipment to list of statutory nuisances and allows local authorities to serve an abatement notice on the person responsible or if the person cannot be found, to attach it to the vehicle.
- Gives local authorities powers to apply various provisions relating to audible intruder alarms to their area.
- Amends and extends EPA 1990 Part III in relation to recovering the expenses reasonably incurred in abating a statutory nuisance from the owner of premises. The Act provides that where such expenses are recoverable, there will be a charge on the premises which is payable in instalments which will include a reasonable rate of interest set by the authority.

REGULATORY AUTHORITY

- Local authority.

COVERAGE

- Noise that is prejudicial to health or a nuisance and is emitted from or caused by a vehicle, machinery or equipment in a street (excluding traffic noise or noise from political demonstrations).
- Audible intruder alarms.

COMPLIANCE REQUIREMENTS

- Where the local authority is satisfied that noise from vehicles, machinery or equipment in the street is causing, or is likely to cause a nuisance, it must serve an abatement notice on the person responsible or if the person cannot be found, attach it to the vehicle. Persons served with notice must comply with its terms, which can:
  - Prohibit or restrict the occurrence or re-occurrence of the noise nuisance.
  - Require the execution of any necessary work to abate the nuisance.
- Person authorised by the local authority can enter or open the vehicle or remove it to a secure place for the purpose of abating the nuisance.
- Where local authority has implemented the provisions, operators of premises must operate audible intruder alarms in accordance with various requirements.

DISCLOSURE

COMPLEMENTARY LEGISLATION

- Control of Pollution Act 1974.
- Environmental Protection Act 1990 Part III.
- Statutory Nuisance (Appeals) Regulations 1995.

RELATED EC LEGISLATION

COMMENTS

RELATED DOCUMENTATION

- Any abatement notice served.
- Any charging notice.

12.3 Outdoor Equipment

LEGISLATION

- Noise Emission in the Environment by Equipment for use Outdoors Regulations 2001 (SI 2001 No. 1701).
- Noise Emission in the Environment by Equipment for use Outdoors (Amendment) Regulations 2001 (SI 2001 No. 3958).

KEY DATES

- In force from 3/7/01 except regulations concerning citation, commencement, revocation, interpretation and appointment of notified bodies which were in force from 4/6/01.
- Amendment Regulations in force from 3/1/02.

PURPOSE

- Implement EC Directive 2000/14 on the approximation of the laws of the Member States relating to the noise emission in the environment by equipment for use outdoors.
- To ensure equipment for use outdoors does not create excessive noise.

APPROACH

- Equipment for use outdoors must satisfy requirements concerning the noise emission in the environment.

REGULATORY AUTHORITY

- Secretary of State.
- Vehicle Certification Agency (an Executive Agency of the Department of the Environment).

COVERAGE

- Outdoor equipment as specified in Schedules 1 and 2 including compressors, hydraulic or rope-operated excavators, landfill compactors, mobile cranes and brush cutters. Schedule 4 defines each type of equipment.
- Regulations do not apply to equipment placed on the market before 3rd July 2001.
- Exemptions exist for equipment put into service outside the Community and for equipment exhibited at trade fairs and exhibitions.

COMPLIANCE REQUIREMENTS

- Must not place on the market or put into service any outdoor equipment unless the following are complied with:
  - An appropriate conformity assessment procedure has been completed.
  - It bears the CE (European Approved Standard) mark and the indication of guaranteed sound power level.
  - It is accompanied by an EC declaration of conformity. The minimum contents of this declaration must be as specified.
  - The guaranteed sound power level does not exceed the applicable permissible sound power level as specified and measured in accordance with the Regulations.
  - The guaranteed sound power level shall be marked on specified equipment.
  - A copy of the declaration of conformity must be sent to the Department of Trade and Industry and the European Commission.
- Must present a copy of technical documentation to a notified body. The notified body will then either require modifications to the technical documentation, or tests, or confirm that the technical documentation satisfies the EC Directive. Notified bodies are ones appointed by the Secretary of State to carry out one, or more, of the conformity assessment procedures.
- Must pay fees.

DISCLOSURE

- Declaration of conformity.
- Technical documentation.

COMPLEMENTARY LEGISLATION

RELATED EC LEGISLATION

- EC Directive 2000/14 on the approximation of the laws of the Member States relating to the noise emission in the environment by equipment for use outdoors.

COMMENTS

- Transitional provisions exist for equipment placed on the market on or before 2nd January 2002. Such equipment is excluded from the Regulations provided that they comply with any noise legislation, relating to the placing on the market in the UK of that equipment, which is in force on 2nd July 2001.
- 2001 Amendment Regulations made only minor corrections to the original Regulations.

RELATED DOCUMENTATION

- Test results.
- Communications with notified body.
- Technical documentation.
- Declaration of conformity.

12.4 Road Vehicles

LEGISLATION

- Road Vehicles (Construction and Use) Regulations 1986 (SI 1986 No. 1078) (as amended).

KEY DATES

- In force from 11/8/86.

PURPOSE

- Impose provisions relating to construction, equipment and maintenance of road vehicles and also conditions relating to their use, including noise emission limits.

APPROACH

- Introduce provisions relating to construction, equipment and maintenance of road vehicles and also conditions relating to their use.
- Introduce type approval schemes, which require any manufacturer, before marketing a new type of vehicle in the UK, to have a sample tested by the Department of the Environment, Transport and Regions.
- Introduce permissible noise limits for all motor vehicles.

REGULATORY AUTHORITY

- Department of Environment, Transport and Regions.

COVERAGE

- All road vehicles (from motorcycles to heavy goods vehicles engines).

COMPLIANCE REQUIREMENTS

- Must comply with new noise limits for all new road vehicles with at least three wheels first used on or after 1st October 1996:

Passenger cars	74 dB(A)
Large buses and coaches greater than 3.5 tonnes:	
Engines less than 150 kW	78 dB(A)
Engines more than 150 kW	77 dB(A)
Small buses and light goods vehicles of 3.5 tonnes or less:	
Gross weight of 2 tonnes or less	76 dB(A)
Gross weight of 2 tonnes or more	77 dB(A)
Heavy goods vehicles over 2 tonnes:	
Engines less than 75 kW	77 dB(A)
Engines 75–150 kW	78 dB(A)
Engines 150 kW or more	80 dB(A)

- "Use" provisions relating to noise include requirement that no motor vehicle shall be used on a road so as to cause any excessive noise which could have been avoided by the exercise of reasonable care on the part of the driver.
- Must comply with requirement that driver of vehicles shall stop the engine when the vehicle is stationary to prevent noise and polluting emissions, except in certain specified circumstances.
- Must comply with requirement that audible warning instruments shall not be used on a restricted road between 23.30 hours and 07.00 hours on the following morning.

DISCLOSURE

COMPLEMENTARY LEGISLATION

- Road Traffic Act 1988.

RELATED EC LEGISLATION

- EC Directive 70/157 relating to Permissible Sound Levels and Exhaust Systems of Motor Vehicles as amended by EC Directive 92/97.
- EC Directive 78/1015 relating to Permissible Sound Levels and Exhaust Systems of Motorcycles as amended by EC Directive 89/235.

COMMENTS

RELATED DOCUMENTATION

12.5 Statutory Nuisances

LEGISLATION

- Environmental Protection Act 1990 Part III: Statutory Nuisances and Clean Air.
- Statutory Nuisance (Appeals) Regulations 1995 (SI 1995 No. 2644).

KEY DATES

- In force from 1/1/91.
- 1995 Regulations in force from 8/11/95.

PURPOSE

- Protect human health and prevent/minimise noise in the environment by giving local authorities preventative and remedial powers to deal with noise from factories, shops and other premises (as with statutory nuisances).

APPROACH

- Establishes that noise emitted from premises can be a statutory nuisance (see Coverage), and requires local authorities to inspect their area periodically to detect whether any such nuisances exist.
- Where local authority is satisfied that a statutory nuisance exists, or is likely to occur, it must serve an abatement notice on the responsible person, or the owner of the premises where that person cannot be found or the nuisance has not yet occurred. The abatement notice can prohibit or restrict the occurrence or re-occurrence of the nuisance and require the execution of any necessary work.
- Defence is available that industry, trade or business is using Best Practicable Means (BPM) to prevent, or counteract the effects of, the nuisance.

REGULATORY AUTHORITY

- Local authority.

COVERAGE

- Covers any premises, industrial/trade/business premises, where statutory nuisance exists.
- List of statutory nuisances includes "noise [including vibration] emitted from premises so as to be prejudicial to health or a nuisance".
- Nuisance is not defined by the Act (*i.e.* there is no specified maximum permitted noise level) and is determined by common law in the courts.

COMPLIANCE REQUIREMENTS

- Where abatement notice has been served, the responsible person (or the owner of the premises where that person cannot be found or the nuisance has not yet occurred) must comply with terms of notice, which can:
  - Prohibit or restrict the occurrence or reoccurrence of the noise nuisance.
  - Require the execution of any necessary work to abate the nuisance. Non-compliance is a criminal offence (See Comments).
- Local authority can recover expenses reasonably incurred in abating statutory nuisance from responsible person.

DISCLOSURE

COMPLEMENTARY LEGISLATION

- Control of Pollution Act 1974.
- Noise and Statutory Nuisance Act 1993.
- Statutory Nuisance (Appeals) Regulations 1995.
- Noise Act 1996 (see Comments).

RELATED EC LEGISLATION

COMMENTS

- Grounds of Appeal exists. See further the Statutory Nuisance (Appeals) Regulations 1995. Persons served with the abatement notice may appeal against it to a magistrates court within 21 days.
- Noise Act 1996 creates a new criminal offence, separate to the statutory nuisance regime, of emitting noise from domestic premises at night which exceeds a specified permitted level The operation of the Act is dependent on its adoption by local authorities, although the Secretary of State also has the power to order that it apply within the area of a local authority.

RELATED DOCUMENTATION

- Any abatement notice served.
- Any charging notice.

13 HAZARDOUS MATERIALS

13.1 Cadmium

LEGISLATION

- Environmental Protection (Controls on Injurious Substances) (No. 2) Regulations 1993 (SI 1993 No. 1643).

KEY DATES

- In force from 31/7/93.

PURPOSE

- Control the marketing and use of cadmium.
- Implement – as regards provisions restricting the marketing and use of cadmium – EC Directive 76/769 on Marketing and Use of Certain Dangerous Substances and Preparations as amended for the tenth time by EC Directive 91/338.

APPROACH

- Restrict the use of cadmium to give colour to finished products made from certain plastics, and the marketing of such products.
- Restrict the marketing of certain paints containing cadmium above a certain level.
- Restricts the use of cadmium to stabilise certain finished products made from PVC, and the marketing of such products.
- Restricts the use of cadmium in the metallic plating of certain products, and the marketing of such products.

REGULATORY AUTHORITY

- Secretary of State.

COVERAGE

- Regulations do not apply to marketing or use for research and development or analysis purposes, products containing cadmium already covered by other EC legislation, or finished products either coloured with cadmium or using cadmium-based stabilisers for safety reasons. Other exceptions also exist.

COMPLIANCE REQUIREMENTS

- Comply with various prohibitions and restrictions on marketing and use of cadmium.
- Breaching the Regulations is a criminal offence.

DISCLOSURE

COMPLEMENTARY LEGISLATION

- Environmental Protection Act 1990 S 140.

RELATED EC LEGISLATION

- EC Directive 76/769 on Marketing and Use of Certain Dangerous Substances and Preparations as amended for the tenth time by EC Directive 91/338.

COMMENTS

RELATED DOCUMENTATION

13.2 Hexachloroethane

LEGISLATION

- Environmental Protection (Controls on Hexachloroethane) Regulations 1998 (SI 1998 No. 545).

KEY DATES

- In force from 1/4/98.

PURPOSE

- Implement in England, Wales and Scotland a 1997 EC Directive 97/16 prohibiting the use of hexachloroethane in the manufacturing and processing of non-ferrous metals.

APPROACH

- Sets out general prohibition on the use of hexachloroethane in the manufacturing and processing of non-ferrous metals (subject to certain exceptions – see Coverage).

REGULATORY AUTHORITY

- Secretary of State.

COVERAGE

Regulation 4 provides for various exceptions to the general prohibition including the use of hexachloroethane:

- For Research and Development or analysis purposes.
- In non-integrated aluminium foundries producing specialised castings for applications requiring high quality and safety standards and where consumption is less than 1.5 kg per day.
- For grain refining in the production of the magnesium alloys AZ81, AZ91 and AZ92.

COMPLIANCE REQUIREMENTS

- Comply with prohibition on use of hexachloroethane.
- Breaching the Regulations is a criminal offence.

DISCLOSURE

COMPLEMENTARY LEGISLATION

- Environmental Protection Act 1990 s 140.

RELATED EC LEGISLATION

- EC Directive 97/16 prohibiting the use of hexachloroethane in the manufacturing and processing of non-ferrous metals, which amends for the 15th time EC Directive 76/769 on the approximation of the laws, regulations and administrative provisions of the member states relating to restrictions on the marketing and use of certain dangerous substances and preparations.

COMMENTS

RELATED DOCUMENTATION

13.3 Mercury Compounds/DBB *etc.*

LEGISLATION

- Environmental Protection (Controls on Injurious Substances) Regulations 1992 (SI 1992 No. 31).

KEY DATES

- In force from 28/2/92.

PURPOSE

- Control the marketing and use of certain injurious substances, including lead carbonate and lead sulphate in paint, mercury compounds in heavy duty textiles, mercury, arsenic and organostannic compounds in the treatment of industrial waters.
- Implement – as regards provisions restricting the marketing and use of certain injurious substances – the EC Directive 76/769 on Marketing and Use of Certain Dangerous Substances and Preparations as amended for the eighth time by EC Directive 89/677.

APPROACH

- Prohibit the supply and use of lead carbonate and lead sulphate in paint except for its supply and use, under certain conditions, for the restoration and maintenance of certain historic buildings or of fine or decorative works of art.
- Prohibit the supply and use of mercury compounds in heavy duty textiles.
- Prohibit the supply and use of mercury, arsenic and organostannic compounds for the treatment of industrial waters.
- Prohibit the supply and use of DBB in a concentration equal to or above 0.1% by weight unless it is for conversion into finished products with a concentration below that level.

REGULATORY AUTHORITY

- Secretary of State.
- Historic Buildings and Monuments Commission.
- Museum and Galleries Commission.

COVERAGE

COMPLIANCE REQUIREMENTS

- Comply with restriction on the supply and use of lead carbonate and lead sulphate in paint.
- Where lead carbonate and lead sulphate in paint is to be supplied or used, under certain conditions, for the restoration and maintenance of certain historic buildings or of fine or decorative works of art, supplier and user must complete written declaration to regulatory authority (in the form set out in the Schedule to the Regulations) at least three weeks in advance.
- Comply with prohibitions on the supply and use of mercury compounds in heavy duty textiles, of mercury, arsenic and organostannic compounds for the treatment of industrial waters and of DBB in a concentration equal to or above 0.1% by weight unless it is for conversion into finished products with a concentration below that level.
- Breaching the Regulations is a criminal offence.

DISCLOSURE

COMPLEMENTARY LEGISLATION

- Environmental Protection Act 1990 S 140.
- Control of Pollution (Supply and Use of Injurious Substances) Regulations 1986.

RELATED EC LEGISLATION

- EC Directive 76/769 on Marketing and Use of Certain Dangerous Substances and Preparations as amended for the eighth time by EC Directive 89/677.

COMMENTS

RELATED DOCUMENTATION

- Any written declaration to regulatory authority, in the form set out in the Schedule to the Regulations, relating to supply and use of lead carbonate and lead sulphate in paint.

13.4 PCBs/PCTs

LEGISLATION

- Environmental Protection (Disposal of Polychlorinated Biphenyls and Other Dangerous Substances) (England and Wales) Regulations 2000 (SI 2000 No. 1043).
- Environmental Protection (Disposal of Polychlorinated Biphenyls and Other Dangerous Substances) (England and Wales) (Amendment) Regulations 2000 (SI 2000 No. 3359).

KEY DATES

- In force from 4/5/00.
- Amendment Regulations in force from 1/1/00.

PURPOSE

- Implement provisions of EC Directive 96/59 on the disposal of polychlorinated biphenyls and polychlorinated terphenyls (PCBs) which require decontamination or disposal of PCBs and equipment containing them and various associated provisions for inventories, labelling and monitoring.

APPROACH

- Prohibit the holding of certain equipment contaminated by PCBs ("contaminated equipment" – see Coverage) by anyone other than a person registered with the Environment Agency, subject to exceptions, *e.g.* businesses concerned with decontamination and disposal of PCBs. This prohibition applied after 31st July 2000.
- Prohibit the holding of PCBs and equipment containing them (whether or not within the definition of "contaminated equipment"), subject to a broader number of exceptions (such as businesses concerned with decontamination and disposal of PCBs, for transformers and for certain other equipment being

replaced under major replacement programmes). This provision applies even though a person is registered with the Agency, but applied only after 31st December 2000.
- Sets out various associated provisions for labelling and monitoring.

REGULATORY AUTHORITY

- Environment Agency.
- Secretary of State.

COVERAGE

- "Contaminated equipment" is defined as any equipment, including any transformer, capacitor or receptacle containing residual stocks which (a) contains PCBs; or (b) having contained PCBs, has not been decontaminated, other than one which contains a total volume of PCBs not exceeding 5 dm^3; and for the purposes of this definition – (i) the PCB volume of a power capacitor is measured by reference to all of the separate elements of a combined set, and (ii) any thing of a type which may contain PCBs will be treated as containing PCBs unless it is reasonable to assume the contrary.

COMPLIANCE REQUIREMENTS

- Must comply with prohibitions on holding of PCBs and equipment containing them (see Approach above).
- Contaminated equipment must be labelled as such and, when decontaminated, within specified limits, must be re-labelled in the form set out in Schedule 2 to the Regulations.
- Applicants for registration must specify, amongst other things, the date by which they expect to have decontaminated or disposed of the equipment concerned. Agency can cancel the registration (only on certain specified grounds). Appeals against the non-determination of an application for registration or the cancellation of a registration also exist.

- Certain charges are payable to the Agency under the Environment Act 1995.
- Breach of prohibitions and various other provisions is a criminal offence.

DISCLOSURE

- Registered holders are required to provide information to the Agency on a regular basis.
- Environment Agency are also placed under various duties by the Regulations, to prepare an inventory and to review it at regular intervals, to provide an annual statement of the number of registered holders and items of equipment of which particulars are registered, and in relation to maintenance of public registers. The latter must include copies of all inventories.
- The Agency also obtains other relevant information through the system of consignment notes under the Special Waste Regulations 1996, which must also be made available to the public.

COMPLEMENTARY LEGISLATION

- Environmental Protection Act 1990 S 140.
- Environmental Protection Act 1990 Part II.
- Environmental Information Regulations 1992.
- Environment Act 1995.

RELATED EC LEGISLATION

- EC Directive 96/59 on the disposal of polychlorinated biphenyls and polychlorinated terphenyls.

COMMENTS

- Regulations revoke the Control of Pollution (Supply and Use of Injurious Substances) Regulations 1986 and subsequent amendments.

- Schedule 1 to the Regulations records the disposal operations D8 to D10, D12 and D15 set out in Annex IIA of EC Directive 75/442 on waste, as replaced by Council Decision 96/350. These are the only permitted methods of disposal of PCBs and equipment contaminated by PCBs under the Directive.

RELATED DOCUMENTATION

- Any application for registration.
- Registered holders must maintain various information, including locations and descriptions of contaminated equipment and quantities and dates and types of treatment.

13.5 PCP

LEGISLATION

- Environmental Protection (Controls on Injurious Substances) Regulations 1993 (SI 1993 No. 1).
- Environmental Protection (Controls on Injurious Substances) (Amendment) Regulations 2001 (SI 2001 No. 3141).

KEY DATES

- 1993 Regulations in force from 31/1/93.
- 2001 Regulations in force from 11/10/01.

PURPOSE

- Control the marketing and use of Pentachlorophenol (PCP).
- Implement – as regards provisions restricting the marketing and use of PCP – EC Directive 76/769 on Marketing and Use of Certain Dangerous Substances and Preparations as amended by EC Directives 91/173 and 99/51.

APPROACH

- Prohibit placing on the market any substance containing PCP in concentration equal to or greater than 0.1% by weight, other than under certain conditions for use in industrial installations for the impregnation of fibres or heavy duty textiles or as a synthesising and/or processing agent.
- Prohibit use of any substance containing PCP in concentration equal to or greater than 0.1% by weight, other than for the purposes for which it may be marketed.
- Prohibit use of wood treated with PCP inside buildings or for certain containers.

REGULATORY AUTHORITY

- Secretary of State.

COVERAGE

- Regulations do not apply to marketing or use for research and development or analysis purposes, or marketing or use which is otherwise authorised under the Control of Pesticides Regulations 1986.

COMPLIANCE REQUIREMENTS

- Comply with prohibition on placing on the market and use of any substance containing PCP in concentration equal to or greater than 0.1% by weight, except where the marketing or use takes place under certain conditions in industrial installations for the impregnation of fibres or heavy duty textiles or as a synthesising and/or processing agent.
- Comply with prohibition on use of wood treated with PCP inside buildings or for certain containers.
- Breaching the Regulations is a criminal offence.

DISCLOSURE

COMPLEMENTARY LEGISLATION

- Control of Pesticides Regulations 1986.
- Environmental Protection Act 1990 S 140.

RELATED EC LEGISLATION

- EC Directive 76/769 on Marketing and Use of Certain Dangerous Substances and Preparations as amended EC Directives 91/173 and 99/51.

COMMENTS

- 2001 Amendment Regulations changed conditions for exemptions, on placing on the market and use, to apply. The exemptions relating to fibres or heavy duty textiles were changed from any substance containing PCP having a total H6CDD content from 4 parts per million to 2 parts per million. Also the exemption now only applies until 31st December 2008.

RELATED DOCUMENTATION

13.6 Ugilec 121/Ugilec 141/DBBT

LEGISLATION

- Environmental Protection (Controls on Injurious Substances) (No. 2) Regulations 1992 (SI 1992 No. 1583).

KEY DATES

- In force from 31/7/92.

PURPOSE

- Control the marketing and use of certain injurious substances, including Ugilec 141, Ugilec 121 (also known as Ugilec 21) and DBBT.
- Implement – as regards provisions restricting the marketing and use of Ugilec 141, Ugilec 121 (also known as Ugilec 21) and DBBT – the EC Directive 76/769 on Marketing and Use of Certain Dangerous Substances and Preparations as amended for the eleventh time by EC Directive 91/339.

APPROACH

- Prohibit the marketing and use of Ugilec 141, or any substance or article containing Ugilec 141.
- Prohibit the marketing and use of Ugilec 121 (also known as Ugilec 21) and DBBT, or any substance or article containing Ugilec 121 or Ugilec 21 or DBBT.

REGULATORY AUTHORITY

- Secretary of State.

COVERAGE

- Regulations do not apply to marketing or use for research and development or analysis purposes.

COMPLIANCE REQUIREMENTS

- Comply with prohibition on the marketing and use of Ugilec 141, or any substance or article containing Ugilec 141, except where the marketing of Ugilec 141 is for use in relation to certain plant and machinery already in service on 18/6/94.
- Comply with prohibition on the marketing and use of Ugilec 121 (also known as Ugilec 21) and DBBT, or any substance or article containing Ugilec 121 or Ugilec 21 or DBBT.
- Breaching the Regulations is a criminal offence.

DISCLOSURE

COMPLEMENTARY LEGISLATION

- Environmental Protection Act 1990 S 140.

RELATED EC LEGISLATION

- EC Directive 76/769 on Marketing and Use of Certain Dangerous Substances and Preparations as amended for the eleventh time by Directive 91/339.

COMMENTS

RELATED DOCUMENTATION

14 ENVIRONMENTAL INFORMATION

14.1 Ecolabelling

LEGISLATION

- EC Regulation 880/92 on a Community Ecolabel Award Scheme. Note: This is directly applicable law in the UK.
- United Kingdom Ecolabelling Board (Abolition) Regulations 1999 (SI 1999 No. 931).

KEY DATES

- In force from 23/3/92.
- In force from 19/4/99 (the UK Regulations).

PURPOSE

- Aims to promote the design, production, marketing, and use of products which have a reduced environmental impact during their entire life cycle, while providing consumers with better information on the environmental impacts of products.
- The 1999 UK Regulations abolish the UK Ecolabelling Board, and appoint in its place the Secretary of State, *i.e.* the Department of the Environment, Transport and Regions (DETR) as the Competent Body for the scheme in the UK.

APPROACH

- The scheme operates by setting different ecological criteria which must be met within specific product groups. The competent body can then award ecolabels to products meeting these criteria within the different product groups.
- The scheme is voluntary but it is intended that products with an ecolabel will gain competitive advantage.

REGULATORY AUTHORITY

- Department of the Environment, Transport and Regions.
- European Commission.

COVERAGE

- Regulation applies to all goods except (1) food, drink or pharmaceutical products and (2) products which are classified as dangerous under Dangerous Preparations Directives 67/548/EEC and 88/379/EEC or products manufactured from processes likely to cause significant harm to humans or the environment.
- Product groups chosen so far range from writing paper to soil improvers.

COMPLIANCE REQUIREMENTS

- Apply voluntarily to competent body if manufacturer, or importer from outside the EU, of product covered by the scheme.
- Application fee payable and an annual licence fee if successful.
- Applicant manufacturer must show compliance with ecological criteria for relevant product group. Certification and documents (including the results of independent testing) must be presented to the competent body.
- Successful applicant must enter into standard form contract with competent body setting out terms of use of label.
- Must comply with prohibition on false and misleading advertising. In addition, references to the eco-label in advertising may not be made until label awarded and then only in relation to the specific product for which it was awarded.

DISCLOSURE

- European Commission to maintain registers of applications received, approved, rejected. These are accessible only to the competent bodies of Member States.

- Commission and competent bodies may not disclose information to which they have gained access in the course of assessing a product to third parties.

COMPLEMENTARY LEGISLATION

- Trades Descriptions Act 1968.

RELATED EC LEGISLATION

COMMENTS

- The European Commission is required to publish in the *Official Journal of the European Communities* the product groups, ecological criteria (and their respective periods of validity) and a list of products for which an ecolabel has been awarded.

RELATED DOCUMENTATION

- Applications.
- Certification and documents (including technical information demonstrating compliance with ecological criteria and the results of independent testing).
- Standard form contract.

14.2 Environmental Information

LEGISLATION

- Environmental Information Regulations 1992 (SI 1992 No. 3240).
- Environmental Information (Amendment) Regulations 1998 (SI 1998 No. 1447).

KEY DATES

- In force from 31/12/92.
- In force from 10/7/98.

PURPOSE

- Grant public rights of access to environmental information, beyond that already maintained in registers, which is held by "relevant persons" (*i.e.* public authorities – see Coverage).
- Implement the EC Directive 90/313 of Freedom of Access to Information on the Environment.

APPROACH

- Imposes a duty on "relevant persons" to make available to the public (on request) information which they have in their possession in relation to the environment (with certain exceptions – see Comments).

REGULATORY AUTHORITY

COVERAGE

- Regulations apply to any information which relates to the environment, which includes the state of any water, air, flora, fauna, natural site, or other land; or any activities or measures adversely affecting or designed to protect the environment, held by a relevant person in accessible form where no other statutory obligation to disclose exists.

- "Relevant persons" include ministers, government departments, local authorities, public utilities and certain types of public body with public administration functions or public responsibilities in relation to the environment.

COMPLIANCE REQUIREMENTS

- Relevant persons are required to supply environmental information to every person who requests it within two months and if they refuse, on one of specified grounds, they must specify in writing the reasons for the refusal.
- Regulations allow for charges to be made in connection with the supply of information.

DISCLOSURE

- Information to be available on request, subject to exceptions contained in the 1998 Regulations including firstly information held for purposes of judicial/legislative functions or secondly information capable of being treated as confidential, only if it relates to national defence/international relations, legal proceedings, internal corporate communications, commercial confidentiality or because the records are still in the course of completion, (see Comments).

COMPLEMENTARY LEGISLATION

- Environment and Safety Information Act 1988.
- Other legislation with provisions requiring information to be placed on public registers (see Comments).

RELATED EC LEGISLATION

- EC Directive 90/313 of Freedom of Access to Information on the Environment.

COMMENTS

- Various public registers exist under the following legislation: the Water Resources Act 1991 (register of discharge consents to controlled waters), the Water Industry Act 1991 (register of discharge consents to sewer), the Environmental Protection Act 1990 (register of waste management licences) and the Town and Country Planning Act 1990 (provides for public access to planning registers).
- Environmental Information (Amendment) Regulations 1998 (SI 1998 No.1447) which came into force on 10th July 1998, have amended the 1992 Regulations so as to fully align the exceptions to the requirement imposed on public bodies to disclose environmental information – contained in Regulation 4(b) of those regulations – with the original exemptions set out in the 1990 Directive on the freedom of access to information on the environment. The five categories of information capable of being treated as "confidential" have not substantively altered, but their wording has been made more restrictive.

RELATED DOCUMENTATION

15 PLANNING AND ENVIRONMENTAL ASSESSMENT

15.1 Environment Impact

LEGISLATION

- Town and Country Planning (Environmental Impact Assessment) Regulations 1999 (SI 1999 No. 293).
- Town and Country Planning (Environmental Impact Assessment) (England and Wales) (Amendment) Regulations 2000 (SI 2000 No. 2867).

KEY DATES

- In force from 14/3/99.
- 2000 Regulations in force from 15/10/00.

PURPOSE

- Assess effects of certain public or private projects on environment before development consent granted for them by public authorities. The 1999 Regulations revoke, and re-enact with amendments, the Town and Country Planning (Assessment of Environmental Effects) Regulations 1988 (as well as later instruments amending them and various supporting pieces of legislation).
- Implement the EC Directive 85/337 on the Assessment of the Effects of certain Public and Private Projects on the Environment, and now the amending EC Directive 97/11. Note that the latter, which the 1999 Regulations were introduced to implement, considerably extends the range of projects requiring an assessment and clarifies the procedures to be used when conducting and recording the assessment.

- Note that although the 1999 Regulations are the main legal instrument for implementing EC Directive 85/337 (as amended by EC Directive 97/11) through UK planning law, the EC Environmental Assessment Directives have also been implemented in the UK through a series of parallel regulations which specifically cover developments not normally subject to planning permission (and therefore not controlled under UK planning law), *e.g.* transport (trunk roads and motorways), off-shore oil drilling and fish-farming (see Complementary Legislation).
- 2000 Regulations amend the 1999 Regulations in relation to applications to mineral planning authorities to determine the conditions to which a planning permission is subject.

APPROACH

- Specify types of projects requiring assessment. The 1999 Regulations list projects in Schedule 1 for which the assessment procedure is mandatory, and projects in Schedule 2 for which assessment will be required where they are likely to have "significant effects" on environment (which is determined by the Local Planning Authority (LPA). Under the new 1999 Regulations, the LPA must use a more formal procedure, and the selection criteria contained in Schedule 3, for "screening" Schedule 2 developments and determining whether Environmental Impact Assessment (EIA) is required.
- Specify application procedures. Developer must submit environmental statement with planning application. Failure to supply an environmental statement where one is required means that the planning application cannot be approved or in some cases cannot be considered.
- LPA must collect and consider information on environmental impact of specified projects when determining application for planning permission. Such a decision follows the same route as a normal planning application, except that the LPA has 16 weeks to review application and accompanying statement, instead of usual 8 weeks. Once the final decisions relating to an EIA application have been made by the LPA or Secretary of State,

they must issue a publicly available statement indicating the reasons for their decision and a description, where necessary, of the main measures to avoid, reduce or, if possible, offset the major adverse effects of the development. In the event of a refusal of planning permission or a dispute as to whether assessment is required the applicant has the usual right to appeal to the Secretary of State.

REGULATORY AUTHORITY

- Local Planning Authority (which must inform Secretary of State of decisions).
- Secretary of State.
- Certain other 'statutory consultees', (*e.g.* the Environment Agency, English Nature and the Countryside Commission for England) must also be provided with details of the proposed development.

COVERAGE

- Projects listed in Schedule 1 of Regulations (for which assessment procedure is mandatory).
- Projects listed in Column 1 of Schedule 2 to the 1999 Regulations where:
  - The project is carried out in certain specified 'sensitive areas', *e.g.* Sites of Special Scientific Interest (SSSIs), National Parks, areas of outstanding natural beauty, the Broads, or land subject to a nature conservation order; *or*
  - The project exceeds the thresholds and criteria laid out in Column 2 of Schedule 2.

    For such Schedule 2 developments, an EIA will be required only if the development would be likely to have "significant effects" on the environment, which is determined by LPA on a case-by-case basis (see above).

COMPLIANCE REQUIREMENTS

- Developer to submit environmental statement with planning application to LPA (Persons may apply for a preliminary opinion from LPA as to whether proposed application is subject to Regulations or LPA may notify applicant to carry out an Environmental Statement (ES) where he has not already done so).
- Developers are entitled to seek an opinion from the LPA (a "scoping opinion") or the Secretary of State (a "scoping direction") on the information to be included in an environmental statement. The types of information which may be required are set out in Schedule 4; note that the Environmental Statement (ES) must at the very least include the information prescribed by Part II of Schedule 4:
  - A description of the proposed development, comprising information about the site, the design, size or scale of the development and the activities to be undertaken.
  - Where significant adverse effects are identified, a description of the measures envisaged in order to avoid, reduce or, if possible, remedy those effects.
  - The data necessary to identify and assess the main effects which the proposed development is likely to have on the environment.
  - The main alternatives which the developer considered and the reasons for selecting the development upon which the application is based, taking into account the environmental effects.
  - A summary in non-technical language of the above.
- Where the ES procedure is applied after the application for planning permission has been made, the developer must comply with various publicity requirements – including publishing a notice relating to the development in a local newspaper – before the ES is actually submitted.

DISCLOSURE

- Details of application – including the ES and any screening and scoping opinions – to be included in local planning register.

- Notice of application to be placed in local newspaper and on project site.
- Environmental statement to be publicly available.

COMPLEMENTARY LEGISLATION

- Town and Country Planning Act 1990.
- Town and Country Planning (General Permitted Development) Order 1995 (as amended) (see Comments).
- Other regulations implementing EC Directive 85/337/EEC, which cover developments, *e.g.* trunk roads and motorways, forestry projects and harbour works not normally subject to planning permission and not therefore controlled under the Town and Country Planning Act 1990. Note: Recent legislation in this area includes:
  - The Offshore Petroleum Production and Pipe-lines (Assessment of Environmental Effects) Regulations 1999 (SI 1999 No. 360).
  - The Highways (Assessment of Environmental Effects) Regulations 1999 (SI 1999 No. 369).
  - The Environmental Impact Assessment (Land Drainage Improvement Works) Regulations 1999 (SI 1999 No. 1783).
  - The Public Gas Transporter Pipe-line Works (Environmental Impact Assessment) Regulations 1999 (SI 1999 No. 1672).
  - The Environmental Impact Assessment (Forestry) (England and Wales) Regulations 1999 (SI 1999 No. 2228).
  - The Harbour Works (Environmental Impact Assessment) Regulations 1999 (SI 1999 No. 3445).
  - Nuclear Reactors (Environmental Impact Assessment for Decommissioning) Regulations 1999 (SI 1999 No. 2892).
  - Electricity Works (Environmental Impact Assessment) (England and Wales) Regulations 2000 (SI 2000 No. 1927).
  - Pipe-line Works (Environmental Impact Assessment) Regulations 2000 (SI 2000 No. 1928).
  - Transport and Works (Assessment of Environmental Effects) Regulations 2000 (SI 2000 No. 3199).

RELATED EC LEGISLATION

- EC Directive 85/337 on the Assessment of the Effects of certain Public and Private Projects on the Environment, as amended by EC Directive 97/11.

COMMENTS

- The General Permitted Development Order 1995 provides that where projects, which would otherwise benefit from permitted development rights, are held to require EIA then planning permission is required.

RELATED DOCUMENTATION

- Any opinion from LPA as to whether proposed application is subject to Regulations.
- Any notice from LPA.
- Application for planning permission and any related correspondence.
- Environmental statement.
- Planning consent.
- Buildings, surveyors *etc.* plans.

15.2 Land Use and Development Control

LEGISLATION

- Town and Country Planning Act 1990 (TCPA).
- Town and Country Planning (Use Classes) Order 1987 (SI 1987 No. 764) (as amended).
- Town and Country Planning (Applications) Regulations 1988 (SI 1988 No. 1812).
- Town and Country Planning (General Permitted Development) Order 1995 (SI 1995 No. 418) (as amended).
- Town and Country Planning (General Development Procedure) Order 1995 (SI 1995 No. 419) (as amended).
- Town and Country Planning (Development Plan) Regulations 1999 (SI 1999 No. 3280).

KEY DATES

- In force from 24/5/90 (with certain exceptions).
- 1999 Regulations in force from 4/1/00 (see Comments).

PURPOSE

- Impose controls over land-use and new development.

APPROACH

- Imposes general requirement of planning permission for all acts of development and material changes in land use (Section 57 of the TCPA 1990). There are exceptions to this general rule (developments on Crown Estates land and many agricultural developments do not require planning permission (see Comments). LPA can approve, reject or apply conditions to any development.
- Planning decisions must be consistent with the development plan, which sets the strategy of the individual LPA with regard to the medium and long-term planning goals for its geographic area of responsibility. When deciding an application the LPA

must also have regard to any other material considerations which include the views of interested parties (see Disclosure). In the event of an application being refused, the applicant may appeal to the Secretary of State for the Environment, Transport and Regions (in most cases this will be dealt with by the Planning Inspectorate).

REGULATORY AUTHORITY

- Local Planning Authority.
- Planning Inspectorate.
- Secretary of State.

COVERAGE

- "Development" is defined as the carrying out of building, engineering, mining or other operation in, on, over or under land, or the making of any material change of use of any buildings or land (s 55 of the TCPA 1990).
- The different land "use" classifications are covered in the (Use Classes) Order 1987 (as amended). Four broad bands of use are identified, each subdivided so as to provide 11 classes. In general, a change of use of a building or 'planning unit' within a class will not be considered as an act of development, while a change of use across classes will be subject to planning permission.
- The Town and Country Planning (General Permitted Development) Order 1995 (as amended) gives 'automatic' planning permission for types of development in a number of categories, thus negating the need for individual planning applications. These categories include development work carried out by drainage bodies, sewerage undertakers and local authorities, certain agricultural buildings and operations and certain industrial and warehouse developments.
- Note: Hazardous developments are usually controlled for planning purposes by the Planning (Hazardous Substances) Act 1990 and related regulations.

COMPLIANCE REQUIREMENTS

- Must submit application for development planning permission where so required. Applications for planning permissions must follow the procedures laid down in the (Applications) Regulations 1988 and should include information on the following: siting, design, external appearance, land use, means of access and landscaping.
- Fees in respect of certain applications for planning permission are payable under the Town and Country Planning (Fees for Applications and Deemed Applications) Regulations 1989.
- Where LPA grants conditional approval, applicant must comply with specified conditions (otherwise the proposal will be refused).

DISCLOSURE

- Details of application and any permission granted will be contained on local planning register. Under the procedures of the General Development Procedure Order 1995, the LPA is required to publicise applications and make them available for viewing in the planning register by members of the public and other interested parties.

COMPLEMENTARY LEGISLATION

- Town and Country Planning (Fees for Applications and Deemed Applications) Regulations 1989.
- Planning (Listed Buildings and Conservation Areas) Act 1990.
- Planning (Hazardous Substances) Act 1990.
- Planning and Compensation Act 1991.
- Town and Country Planning General Regulations 1992 (as amended)
- Planning (Control of Major Accident Hazards) Regulations 1999.
- Landfill (England and Wales) Regulations 2002.

RELATED EC LEGISLATION

COMMENTS

- Planning legislation in England and Wales is supported by a series of Department of the Environment, Transport and Regions (England) and Welsh Office (Wales) Circulars and Planning Policy Guidelines (PPGs) which, although not legally binding, are generally accepted as setting the norms for planning applications, decisions and the policies of those required to judge planning applications (in most cases the LPA).
- Reference should be made to *Planning Applications: The RMJM Guide* by Henry Brown and Adrian Salt (1988, 3rd Edition), a useful practical guide to the various procedures for obtaining planning permission containing the statutory forms which have to be served at various stages of an application.
- 1999 Development Plan Regulations prescribe the form and content of structure plans, unitary development plans, local plans, waste local plans and minerals local plans. They prescribe an expanded list of matters to which an LPA must have regard when formulating their plans, and together with the Planning (Control of Major Accident Hazards) Regulations 1999, they serve to implement – in relation to planning in England – Article 12 of EC Directive 96/82 on the control of major accident hazards involving dangerous substances.
- The Regulations include requirements relating to noise.
- Landfill Regulations 2002 require planning authorities to take specified location requirements into consideration when granting planning permission for landfills.

RELATED DOCUMENTATION

- Documentation relating to application for planning permission and any related correspondence.
- Planning consent.
- Any enforcement or stop notice.
- Buildings, surveyors *etc*. plans.

16 NATURE CONSERVATION

16.1 Protection of Wildlife and Countryside

LEGISLATION

- Wildlife and Countryside Act 1981.
- Wildlife and Countryside (Amendment) Act 1991.
- Wildlife and Countryside Act 1981 (Amendment) Regulations 1995 (SI 1995 No. 2825).

KEY DATES

- In force (mostly) from 30/11/81. Other provisions were brought into force on various dates between 16/2/82 and 28/2/83.
- 1991 Act in force from 25/9/91.
- 1995 Regulations in force from 30/11/95.

PURPOSE

- Implement several major reforms to, and form the legal cornerstone of, the UK arrangements for the protection, control and management of endangered wildlife and the countryside which supports it.

APPROACH

- Part I strengthens protection for, and provides "off-site" powers to ensure the conservation of, various endangered species of wildlife including wild birds (transposing in part EC Directive 79/409 on the Conservation of Wild Birds).
- Part II strengthens the legal basis for areas of Sites of Special Scientific Interest (SSSIs) and expands the degree to which the regulatory authorities can restrict the rights of owners and occupiers and influence the management of designated sites.

- Introduces new provisions for the recompense of site owners and occupiers, for both profits foregone and past expenditure on projects aborted in the interests of conservation.
- Part III amends the law relating to public rights of way.

REGULATORY AUTHORITY

- English Nature.
- Countryside Council for Wales.
- Secretary of State.

COVERAGE

- Species to which various provisions of Part I apply are listed in Schedules to 1981 Act.

COMPLIANCE REQUIREMENTS

- Owners and occupiers of any land notified as SSSI must not carry out, or cause or permit to be carried out, any operation specified in notification as likely to damage the flora, fauna or other features unless the relevant regulatory authority has been given written notice.
- Regulatory authority must also give its consent in writing, unless either the operation is in accordance with a Nature Reserve Agreement under the National Parks and Access to the Countryside Act 1949 or a Management Agreement under the Countryside Act 1968 (in which case they are exempt from this requirement as they have already in effect received the authority's approval) or the authority has not responded to the written notice within four months.
- Breach of these requirements without reasonable excuse, which would include an "emergency operation" or the obtaining of planning permission for the operation, constitutes a criminal offence.
- Must comply with various provisions in Part I of 1981 Act relating to the protection of wild birds, animals and plants.

These include prohibitions on certain methods of killing or taking wild birds and animals, provisions protecting certain wild plants and restrictions on the introduction of certain plants and animals to Great Britain.

DISCLOSURE

COMPLEMENTARY LEGISLATION

- National Parks and Access to the Countryside Act 1949.
- Countryside Act 1968.
- Endangered Species (Import and Export) Act 1976.
- Wildlife and Countryside (Service of Notices) Act 1985.
- EPA 1990.
- Conservation (Natural Habitats *etc.*) Regulations 1994 (see Comments).
- Conservation (Natural Habitats *etc.*) (Amendments) Regulations 2000.
- Numerous Regulations enacted to amend the Schedules and thereby uupdate the lists of protected species.

RELATED EC LEGISLATION

- EC Directive 79/409 on the Conservation of Wild Birds.
- EC Directive 92/43 on the Conservation of Natural Habitats and of Wild Fauna and Flora.

COMMENTS

- Detailed advice on the powers under the 1981 Act and their relationship to land-use planning is contained in DoE Circular 27/87.
- For general guidance on planning policy in this area, see also PPG9 on Nature Conservation (published in October 1994).
- The Conservation (Nature Habitat) Regulations 1994 (as amended) implement EC Directives 79/409 (the Birds Directive) and 92/43 (the Habitats Directive) and establish rules for the

protection of Special Protection Areas (SPAs) and Special Areas of Conservation (SACs) so as to conserve natural habitats and wild flora and fauna.

- 1991 Amendment Act made it an offence knowingly to cause or permit to be done certain acts.

RELATED DOCUMENTATION

- Any written notice to relevant regulatory authority.
- Consent from relevant regulatory authority.
- Any relevant planning permission.
- Land details of species *etc*.

17 ENERGY CONSERVATION

17.1 Energy Information – Certain Appliances

LEGISLATION

- Energy Information (Refrigerators and Freezers) Regulations 1994 (SI 1994 No. 3076).
- Energy Information (Washing Machines) Regulations 1996 (SI 1996 No. 600), Energy Information (Washing Machines) (Amendment) Regulations 1997 (SI 1997 No. 803) (see Comments).
- Energy Information (Tumble Driers) Regulations 1996 (SI 1996 No. 601).
- Energy Information (Combined Washer–Driers) Regulations 1997 (SI 1997 No. 1624).
- Energy Information (Lamps) Regulations 1999 (SI 1999 No. 1517).
- Energy Information (Dishwashers) Regulations 1999 (SI 1999 No. 1676).
- Energy Information and Energy Efficiency (Miscellaneous Amendments) Regulations 2001 (SI 2001 No. 3142).

KEY DATES

- In force from 1/1/95 (1994 Regulations).
- In force from 1/5/96 (1996 Regulations).
- In force from 1/8/97 (1997 Regulations) although the requirements do not apply until 1/2/98.
- In force from 1/7/99 and 8/7/99 (1999 Regulations).
- In force from 19/10/01 (2001 Regulations).

PURPOSE

- Encourage energy conservation by providing information to consumers on energy consumption of refrigerators and freezers, washing machines, tumble driers, combined washer–driers, lamps and dishwashers.

APPROACH

- Suppliers of refrigerators and freezers, washing machines, tumble driers, combined washer–driers, lamps and dishwashers must supply labels and information notices to dealers containing information about their energy consumption and placing the appliance in one of seven energy efficiency categories. Other information on operating efficiency, water consumption and noise levels must also be included.
- Dealers must then attach the label to appliance displayed to consumers.

REGULATORY AUTHORITY

- Local Weights and Measures Authority.

COVERAGE

- Refrigerators, frozen food storage cabinets and food freezers, washing machines (with internal means to heat water – see Comments), tumble driers, combined washer–driers, and dishwashers which are electric mains operated and unable to use other energy sources. Regulations do not apply to second hand appliances or appliances for which production ceased before the date when the regulation concerned came into force.
- Lamp regulations apply to household electric lamps supplied from mains and household fluorescent lamps.

COMPLIANCE REQUIREMENTS

- Suppliers must supply labels and information notices to dealers containing information about the energy consumption of refrigerators and freezers, washing machines, tumble driers, combined washer–driers, lamps and dishwashers.
- Suppliers are responsible for the accuracy of the information and must establish technical documentation enabling it to be assessed.
- Enforcement authorities have various powers to require technical documentation from suppliers where they believe

information to be incorrect and enter premises and inspect, seize and detain appliances.

- Dealers must attach the label to appliance displayed to consumers.

DISCLOSURE

- Supplier must make the technical documentation available for inspection by enforcement authorities for five years after the appliance has ceased to be manufactured and furnish it promptly to the authorities when so required.

COMPLEMENTARY LEGISLATION

RELATED EC LEGISLATION

- EC Directive 92/75 and EC Directive 94/2 on energy labelling of refrigerators and freezers, EC Directive 95/12 on energy labelling of household washing machines, EC Directive 95/13 on energy labelling of electric tumble driers, EC Directive 96/60 on energy labelling of combined washer–driers, EC Directive 98/11 on energy labelling of household lamps and EC Directives 97/17 and 99/9 on energy labelling of dishwashers.

COMMENTS

- Non-compliance with most of the Regulations' provisions is a criminal offence punishable by a fine.
- Amending Regulations 1997 exclude washing machines with no internal means to heat water from the provisions of the Energy Information (Washing Machines) Regulations 1996.
- Amendment Regulations 2001 removed the requirement for a dealer, when the appliance was not on display, to make an information notice available to customers before concluding a sale. It also made a number of changes to enforcement procedures and drafting amendments.

RELATED DOCUMENTATION

- Labels and information notices.
- Technical documentation enabling information notice/labels to be assessed.

17.2 Building – Energy Usage

LEGISLATION

- Building Regulations 2000 (SI 2000 No. 2531).
- Building (Amendment) Regulations 2001 (SI 2001 No. 3335).

KEY DATES

- In force from 1/1/01.
- 2001 Regulations in force from 1/4/02.

PURPOSE

- Impose requirements on persons carrying out certain building operations. The new 2000 Regulations, made under the Building Act 1984, revoke and replace with amendments the Building Regulations 1991 and consolidate with amendments all subsequent amendments to those Regulations.

APPROACH

- Impose requirements on persons carrying out certain "building work" (as defined – see Coverage) and these requirements include those relating to the conservation of fuel and power (Schedule 1, Part L).
- Local authorities may dispense with or relax any requirement contained in the Regulations.

REGULATORY AUTHORITY

- Local authority.

COVERAGE

- "Building work" includes the erection or extension of a building, the provision or extension of a controlled service or fitting in or in connection with a building; the material alteration of a

building; work required in relation to a material change of use, the insertion of insulating material into the cavity wall of a building; and work involving the underpinning of a building.
- "Building" refers to any permanent or temporary building but not any other kind of structure or erection, and includes a part of a building.

COMPLIANCE REQUIREMENTS

- Where building work (as defined) is being carried out on dwellings, reasonable provision must be made for the conservation of fuel and power by:
  - Limiting the heat loss through the fabric of the building, from hot water pipes and hot air ducts and from hot water vessels.
  - Providing space-heating and hot water systems which are energy efficient.
  - Providing lighting systems with appropriate lamps and sufficient controls so that energy can be used efficiently (only applies to external lighting systems fixed to the building).
  - Provide sufficient information with the heating and hot water services so that occupiers can use energy efficiently.
- For buildings other than dwellings the following additional requirements apply as well. Reasonable provision must be made for the conservation of fuel and power by:
  - Limiting the heat loss through gains as well as losses through the fabric of the building, from hot water service pipes and limiting exposure to solar overheating.
  - Make sure no more energy is used in air conditioning than is reasonable.
  - Limit the heat gains by chilled water and refrigerant vessels and pipes and air ducts that serve air conditioning.
  - Provide light systems which are energy-efficient.
  - Provide information to relevant services rather than just heating and hot water services.
- Schedule J provides requirements for combustion appliances and fuel storage systems for prevention of overheating, reducing risks of burns to people, protection of liquid fuel storage systems from fire, protection against pollution and provision of

information. This includes oil storage tanks and the pipes connecting them to combustion appliances shall be constructed and protected to reduce to a reasonable level the risk of oil escaping and causing pollution. A notice shall be fixed to them, in a prominent position, containing information on how to respond to an oil escape so as to reduce the risk of pollution.

- Various other provisions exist for the conservation of fuel and power in buildings, including the mandatory carrying out of a energy rating for new buildings, but these provisions apply to residential dwellings only.

DISCLOSURE

COMPLEMENTARY LEGISLATION

- Building Act 1984.
- Home Energy Conservation Act 1995.
- Energy Conservation Act 1996.
- Home Energy Efficiency Scheme Regulations 1997.
- Building (Amendment) Regulations 2002.

RELATED EC LEGISLATION

COMMENTS

- 2001 Regulations replaced Schedules J and L from the original 2000 Regulations.
- 2002 Amendment Regulations in force from 1st April 2002 makes minor amendments and adds provisions for replacement windows *etc*.

RELATED DOCUMENTATION

SUBJECT INDEX

USEFUL GUIDES CITED IN THE TEXT
(This is not an exhaustive list)

CHEMICAL ABBREVIATIONS

CFC chlorofluorocarbons
DBB di-+-oxo-di-*n*-butylstanniohydroxyborane ($C_8H_{19}BO_3Sn$)
DBBT monomethyldibromodiphenyl methane
DDT dichlorodiphenyltrichloro ethane
PCB polychlorinated biphenyls
PCP pentachlorophenol
PCT polychlorinated terphenyls
Ugilec 121, Ugilec 21 monomethyldichlorodiphenyl methane
Ugilec 141 monomethyltetrachlorodiphenyl methane